Geesey

ADHESION IN BIOLOGICAL SYSTEMS

CONTRIBUTORS

ROBERT E. BAIER
H. BARNES
SURINDAR N. BHASKAR
JAMES W. BOTHWELL
MICHAEL G. BUONOCORE
JAMES L. CHEN
M. COOK
C. WILLIAM COOPER
WILLIAM A. CORPE
GILMAN N. CYR
RICHARD D. FALB
JOHN D. GALLIGAN
GERALD A. GRODE
ROBERT E. HILLMAN
HENRY L. LEE, JR.
FRED LEONARD
VINCENT R. LIGUORI
JACKSON LYNN
TERUO MATSUMOTO
FRANCIS W. MINOR
PAUL F. NACE
ANTHONY M. SCHWARTZ
MICHAEL L. SWARTZ
A. CECIL TAYLOR
LEONARD WEISS

ADHESION IN BIOLOGICAL SYSTEMS

Edited by RICHARD S. MANLY

School of Dental Medicine
Tufts University
Boston, Massachusetts

1970

ACADEMIC PRESS New York and London

ACADEMIC PRESS, INC.
111 Fifth Avenue, New York, New York 10003

United Kingdom Edition published by
ACADEMIC PRESS, INC. (LONDON) LTD.
Berkeley Square House, London W1X 6BA

LIBRARY OF CONGRESS CATALOG CARD NUMBER: 73-107568

PRINTED IN THE UNITED STATES OF AMERICA

CONTENTS

Applications of Alpha Cyanoacrylates

PART IV. ADHESIVES FOR HARD TISSUES

15. Bonding to Hard Dental Tissues

MICHAEL G. BUONOCORE

16. Polyurethanes

JOHN D. GALLIGAN, FRANCIS W. MINOR, AND ANTHONY M. SCHWARTZ

17. Surface Preparation and Various Adhesive Resins

HENRY L. LEE, JR., AND MICHAEL L. SWARTZ

LIST OF CONTRIBUTORS

Numbers in parentheses indicate the pages on which the authors' contributions begin.

ROBERT E. BAIER, Cornell Aeronautical Laboratory of Cornell, Buffalo, New York (15)

H. BARNES, The Dunstaffnage Marine Research Laboratories, Oban, Scotland (89)

SURINDAR N. BHASKAR, Division of Oral Pathology, United States Army Institute of Dental Research, Walter Reed Army Medical Center, Washington, D. C. (201)

JAMES W. BOTHWELL, Johnson & Johnson Research, New Brunswick, New Jersey (215)

MICHAEL G. BUONOCORE, Eastman Dental Center, Rochester, New York (225)

JAMES L. CHEN, Squibb Institute for Medical Research, New Brunswick, New Jersey (163)

M. COOK, Parker Brewis Research Fellow, The Dental School, University of Newcastle, Newcastle upon Tyne, England (139)

C. WILLIAM COOPER, Battelle Memorial Institute, Columbus Laboratories, Columbus, Ohio (153)

WILLIAM A. CORPE, Department of Biological Sciences, Barnard College, Columbia University, New York, New York (73)

GILMAN N. CYR, Squibb Institute for Medical Research, New Brunswick, New Jersey (163)

RICHARD D. FALB, Battelle Memorial Institute, Columbus Laboratories, Columbus, Ohio (153)

JOHN D. GALLIGAN, Gillete Research Institute, Washington, D. C. (255)

GERALD A. GRODE, Battelle Memorial Institute, Columbus Laboratories, Columbus, Ohio (153)

ROBERT E. HILLMAN, William F. Clapp Laboratory of Battelle Memorial Institute, Duxbury, Massachusetts (113)

HENRY L. LEE, JR., Research and Development Center, The Epoxylite Corporation, South El Monte, California (269)

FRED LEONARD, U. S. Army Biomedical Research Laboratories, Walter Reed Army Medical Center, Washington, D. C. (185)

VINCENT R. LIGUORI, Osborn Laboratory of Marine Sciences, New York Aquarium, Brooklyn, New York (123)

JACKSON LYNN, Battelle Memorial Institute, Columbus Laboratories, Columbus, Ohio (153)

TERUO MATSUMOTO, Department of Surgery, Section of Experimental Surgery, Hahnemann Medical College and Hospital, Philadelphia, Pennsylvania (209)

FRANCIS W. MINOR, Gillette Research Institute, Washington, D. C. (255)

PAUL F. NACE, Department of Biology, Purdue University, Indianapolis, Indiana (113)

ANTHONY M. SCHWARTZ, Gillette Research Institute, Washington, D. C. (255)

MICHAEL L. SWARTZ, Research and Development Center, The Epoxylite Corporation, South El Monte, California (269)

A. CECIL TAYLOR, University of Texas, Dental Branch, Houston, Texas (51)

LEONARD WEISS, Department of Experimental Pathology, Roswell Park Memorial Institute, Buffalo, New York (1)

PREFACE

Several excellent treatises on adhesion have been published in the past, but the emphasis to date has been on types of adhesives commonly used to bond surfaces of paper, wood, or metal. "Adhesion in Biological Systems" is an attempt to bring together a summary of our knowledge of adhesion in the presence of moisture, a condition required in almost all biological systems.

This volume is the outgrowth of an AAAS Symposium held in December of 1967. Since that time, the authors of the various chapters were invited to update their contributions in line with the more recent knowledge available. None of the material, in its present form, has been published elsewhere. Of the 12 papers presented at the symposium, 11 appear in this work. In addition, Drs. Barnes, Bothwell, Buonocore, Cook, Corpe, and Liguori contributed papers to help broaden the coverage of the publication.

It is hoped that the scientists in a number of fields, including physics, chemistry, zoology, botany, engineering, medicine, and pharmacy, will benefit from this book.

As organizer of the symposium and editor of the volume, I would like to acknowledge at least some of the many persons who helped with the project. Dean Scott of Western Reserve, Dean Sognnaes of UCLA, Dr. Chavoor of Georgetown, Dr. Lunin of Maryland, Dr. Johansen of Rochester, Dr. Likins, then of NIH, and Dr. Pearlman, then of the American Dental Association in Chicago, were all members of the Section Nd Advisory Committee. Dr. Kenneth Manly was very helpful in reaching some decisions regarding relevance of parts of various manuscripts. Mrs. Frances Manly did an outstanding job in assisting me in the preparation of the chapters for publication.

PART I

PRINCIPLES OF ADHESION IN BIOLOGICAL SYSTEMS

CHAPTER 1

A BIOPHYSICAL CONSIDERATION OF CELL CONTACT PHENOMENA

*Leonard Weiss**

Cell contact phenomena may be divided into three basic processes, namely, (1) cell contact, (2) cell adhesion and attachment, and (3) cell separation.

I shall first discuss the three processes separately from an abstract, physical approach and then consider the inherent difficulties in applying a simple physical approach to real cellular systems.

I. Cell Contact

All vertebrate cells so far examined carry a net negative surface charge. Contact between cells or their processes and cellular or non-cellular substrata has been considered as analogous to contact between negatively charged particles in suspension (Curtis, 1962; Pethica, 1961; Weiss and Woodbridge, 1967; Brooks *et al.*, 1967; Weiss, 1967a). This approach, based on the theory of lyophobic colloids developed by Derjaguin and Landau (1941), and independently by Verwey and Overbeek (1948), is known as the DLVO-theory. In essence, the theory considers contact to be determined by the balance of elec-

*Some of my own work described here was partially supported by Grant No. P-403A from the American Cancer Society, and Contract No. 45181 from the National Institute for Dental Research.

trostatic forces of repulsion which tend to keep the charged particles apart, and the attractive London-van der Waals' forces which tend to bring them together.

In the case of two similar spherical particles, the potential energy of repulsion V_R, which prevents contact, may be given by:

$$V_R = 1/2\ Dr\ \chi_0^2 \log_e (1 + e^{-Kd}) \qquad (1)$$

where D = dielectric constant of the medium, r = the particle radius of curvature, χ_0 = surface potential, $1/K$ = Debye-Hückel parameter, and d = distance between particle surfaces.

The potential energy of attraction V_A between two solid spheres has been given by Hamaker (1937) as:

$$V_A = \frac{-Ar}{12d} \qquad (2)$$

where A is constant and $2r \gg d$.

In the case of a sphere approaching a flat plate, which may be more relevant to a cell process with a hemispherical end approaching a noncellular surface, the net repulsive barrier V is given by:

$$V = 16Dr\left(\frac{kT}{\epsilon}\gamma\right)^2 \exp(-Kd) - \frac{Ar}{6d} \qquad (3)$$

where $\gamma = [\exp(y)-1]/[\exp(y)+1]$, $y = \epsilon\chi/2kT$, k = Boltzmann's constant, ϵ = electronic charge, and T = absolute temperature

The Debye-Hückel reciprocal length, K, which defines the extent of the diffuse layer of counter ions at a charged interface, is given by:

$$K = \left(\frac{8\pi n\epsilon^2 Z^2}{kTD}\right)^{1/2} \qquad (4)$$

where n = number of ions of each species per unit volume of solution, and Z = valency.

Factors increasing V_R or decreasing V_A will tend to promote contact between cells. Thus, if the equations are relevant, examination of their various component terms should help in a qualitative assessment of contact problems, and enumeration of the terms should provide a quantitative approach.

Present difficulties in the formulation of the equation for V_A make calculations based on it of very doubtful accuracy, and it is not proposed to deal with attractive interactions further. The biophysical aspects of Eq. (2), together with its shortcomings, are discussed at some length elsewhere (Weiss, 1967a). It can be seen from Eqs. (1) and

(3) that cell contact is expected to be promoted by low surface potential (χ_0), by low radius of curvature (r) of contacting cell surfaces, by low environmental dielectric constant (D), and by factors reducing the effective thickness of the electrical double layer of ions associated with the cell surface ($1/K$). Equation (4) shows that $1/K$ may be reduced by raising environmental ionic strength and/or valency.

At this stage, it may be asked how the various terms can be evaluated by experiment on biological material. Surface potential, χ_0, cannot be determined directly, but it may be estimated from measurements of cellular electrophoretic mobility, μ. Where ζ is the zeta potential at the shear plane surrounding a moving particle, the Smoluchowski equation gives:

$$\mu = \frac{D\zeta}{4\pi\eta} \tag{5}$$

where μ is expressed in $cm^2\ volt^{-1}\ second^{-1}$, and η = dynamic viscosity of the suspending fluid at the shear plane. Haydon (1961) has suggested that for values of ζ less than -50 mV, as in the case of mammalian cells, it is reasonable to equate ζ with surface potential.

It cannot be assumed that the surface potential of living cells remains constant throughout their life, or that this parameter is not changed during measurement. Mayhew (1966, 1967) has demonstrated changes in surface charge density and in the nature of surface ionogenic groups related to the mitotic cycle in a number of cells, and Eisenberg, Ben-Or, and Doljanski (1964) have related changes in electrophoretic mobility to regeneration and growth. When renewal of oral epithelium occurs, it is conceivable that the surface charge density of mitosing gingival cells increases above that of those in interphase, and that this affects their contact reactions with tooth elements. In electrokinetic analytical procedures, in which the electrophoretic mobilities of cells are measured in suspending fluids of different ionic strength and pH (Bangham and Pethica, 1960), it has been shown that some of the fluids themselves produce overt changes in cell morphology which may well produce artifactual changes at their surfaces, thereby affecting the validity of the measurements (Pulvertaft and Weiss, 1963). Weiss (1966a) has also demonstrated a temperature-dependent change in cellular electrophoretic mobility which adds weight to the suggestion that cells are best studied in their natural environment. Unfortunately our lack of knowledge of the nature of the normal tissue fluids prevents this condition from being met.

Another problem which runs throughout the whole approach to cell contact phenomena is that many of our experimental techniques are only applicable to single cells, which in turn necessitates the isolation of cells from tissues by enzymic, mechanical, or other types of dissociation procedures. Such isolation may well alter the nature of the cell periphery (Weiss, 1967b). Enzyme treatment, for example, has been shown to affect some cellular contact and adhesive interactions *in vitro* (Weiss and Kapes, 1966).

Quite apart from changes in cell surface charge accompanying division or growth, this parameter may be significantly altered by the action of enzymes which effectively cleave ionized groups from the cell periphery. Among the enzymes in the dentogingival region which could reduce surface charge in this manner are the proteolytic enzymes (Lisanti, 1960) and neuraminidase (Perlitsh and Glickman, 1966). From analogous work on other cells, some of which is reviewed in the symposium edited by Ambrose (1965), it would be surprising if ionogenic groups could not be enzymically cleaved from the electrokinetic surfaces of the cells of the gingival epithelium. Preliminary work by Neiders and Weiss (1968, unpublished) has in fact shown that the electrophoretic mobilities of isolated cells from human gingival epithelium are significantly reduced by treatment with neuraminidase, presumably as a result of the enzyme induced removal of the ionized carboxyl groups of terminal sialic acid moieties from the cell surface. It should be borne in mind that in inflammatory and other lesions, enzymes present in exudates may cause modification of the various parameters affecting cell contact.

Radius of curvature, r, can affect cell contact in at least two ways. Equation (1) shows that diminishing radius of curvature of approaching cell processes reduces the potential energy barrier to contact (Bangham and Pethica, 1960; Pethica, 1961), whereas Eq. (2) shows that this change also reduces the attractive potential which promotes contact. It would appear that reduction in radius of curvature would facilitate very close approach of cells "through" the total potential energy of interaction barrier. Measurements of cell deformability might be made by estimating the negative pressure required to suck a hemispherical bulge from cells into 5-μ micropipettes. A series of studies has indicated that at least some cells may be more easily deformed following treatment with neuraminidase (Weiss, 1965a), proteolytic enzymes (Weiss, 1966b) and calcium chelation (Weiss, 1967c,d).

The effect on contact of dielectric constant D is also indicated by Eq. (1) which shows that increase in the value of this parameter increases the potential energy of repulsion and thereby hinders contact. Attempts to evaluate D in various biological media and to relate the information to agglutination of erythrocytes have been made by Pollack *et al.* (1965), and this work, together with fresh discussion and computations, is given by Brooks *et al.* (1967). However, as noted by Weiss and Woodbridge (1967), if, as may be reasonably expected, the peripheral regions of cells have a different dielectric constant from that of their environment, then at close cellular approach instead of one value for D, three values should be used: one value for the bulk phase, and one for each of the surfaces making contact. At the moment, no precise estimates of the dielectric constant are available.

The Debye-Hückel parameter, $1/K$, which has the dimension of length, will affect contact as indicated in Eq. (1). The effect of raising the ionic strength of the medium is well documented in classic colloidal systems (Verwey and Overbeek, 1948) and has been demonstrated in the case of the adsorption of virus particles (Valentine and Allison, 1959) and the agglutination of leukocytes (Wilkins *et al.*, 1962). It must be remembered that in the physiological situation the ionic strength approximates to 0.15 gm-ions per liter, which is high by "colloid" standards, and that comparatively minor fluctuations are unlikely to affect cell contact directly.

Quite apart from the theoretical problems associated with colloid stability, which have been recently discussed (*Discussions of the Faraday Society*, No. 42, 1966), there are many practical problems standing in the way of using the DLVO approach to problems of cell contact. Observations of cell contact *in vitro* reveal quite clearly that actively moving cells do not simply collide with each other. Rather, they put out probes and explore surfaces as though seeking suitable areas at which to make contact. This has been particularly well documented by the time-lapse movies of P. Weiss and A. C. Taylor among others (see P. Weiss, 1961, for descriptions). Ambrose's experiments (1961) with the surface contact microscope show that a cell cultured on glass, for example, appears to be in true contact with the glass over only a small proportion of the total area in apparent contact. Observations of this type suggest that specific areas of the cell periphery—that is to say parts of a heterogeneous surface—are involved in contact phenomena. Most calculations, however, are made on the entirely unwarrantable assumption of homogeneous cell surfaces. The practical problem is that in our experiments most, if not all, parameters measured give average values. For example, measurements of cellular electrophoretic mobility lead to estimates of net surface charge, whereas the information sought may well relate to charge distribution, which is not obtained by this technique (Weiss, 1963a, 1968a, b). Does cell contact occur by the apposition of charged peripheral regions, or by contact of these regions with uncharged regions? Or, put in different terms, does cell surface charge determine where contact is made, or where it is not made? In my opinion, a systematic, biologically oriented approach to the role of charge in cell contact phenomena has hardly begun at either the theoretical or experimental level.

II. Cell Adhesion and Attachment

When cell contact is made and opposing regions of two cell peripheries are brought close enough together, then adhesion can occur between them. The word "close" can be defined in this context in relation to the various forces of adhesion involved.

Forces which are of possible importance in cell adhesion have been reviewed by Pethica (1961) and Weiss (1967a) and will only be briefly summarized here. They include: (1) chemical bonds (Pauling, 1960) which may be classified as: (a) electrostatic bonds of which the most important are ionic bonds due to coulombic attractions between ions of opposite net charge. Among this group may be included coordinates of the type tentatively proposed by Rappaport and Howze (1966) and Rappaport (1966), in which negatively charged groups at the cell periphery are coordinated about monovalent cations, particularly potassium; (b) covalent bonds; and (c) hydrogen bonds, in which an atom of hydrogen is shared by two strongly electronegative atoms.

All of the chemical bonds described act over approximately 2-3 Å. (2) van der Waals' interactions of the London type are dependent on charge fluctuations which, by causing rapidly changing fields due to change in dipole moments, polarize other molecules, leading to attraction. The nonpolar parts of protein, for example, tend to orient away from an aqueous phase, and to cohere by means of nonpolar, London-van der Waals' interactions. This has been described as the "hydrophobic" bond by Kauzmann (1959). Interactions of this type are expected to be effective over distances of 3–4 Å.

If adhesion is to occur between two cell peripheries by means of the forces described, the interacting regions must come to within about 4 Å of each other when they make contact. At the moment there are no methods available for measuring cell separation of this order of magnitude, or indeed for defining the limits of the cell periphery at this level. The light microscope cannot give the requisite resolution, and the various preparative artifacts and limitations in staining techniques associated with electron microscopy make it unsuitable for approaching this problem. Thus, merely looking at static preparations of cells will not tell us whether contact has taken place, and whether the failure of one cell to adhere to another is due primarily to their failure to make contact—i.e., to approach more closely than 4 Å—or whether effective failure to adhere is unconnected with the contact process.

Adhesion as I have described it occurs at rates governing the formation of the various adhesive bonds, and for all practical purposes may be regarded as instantaneous or short-term. As cells are dynamic systems, longer term adhesive interactions may also occur between two adherends involving the synthesis of cell surface material and structures such as fibers, desmosomes, and hemidesmosomes. Desmosomes and hemidesmosomes develop at sites of attachments between cells and their substrata, which may be cellular or noncellular. Although desmosomes disappear in viable cells following separation from their substrata, this takes a few hours; thus, the appearance of desmosomal structures in isolated cells may be taken to indicate past attachment in this region.

One of the main difficulties in the way of postulating specific adhesive mechanisms involving the cell periphery is our comparative ignorance of the physicochemical nature of this region. In the case of the erythrocyte where membranes may be isolated with minimum trauma, a great deal of analytical data is available, as, for example, in the text edited by Bishop and Surgenor (1964). However, even recent methods for isolating the membranes of tissue cells, which represent considerable advances over earlier techniques (Wallach, 1967; Warren *et al.*, 1967), are at least potentially traumatic and, of course, straightforward chemical analyses do not permit the identified components to be spatially located on specific regions of the cell periphery.

III. Cell Separation

It is well known in the field of adhesives technology (Bikerman, 1961) that if two adherends A and B are separated, the separation seldom if ever occurs at the adhesion interface, A/B. As shown diagrammatically in Fig. 1, ruptures, which may be at a microscopic level, occur within the two adherends because of cohesive failure resulting, after separation, in pieces of A being left on surface B, if B is stronger or more coherent than A, and vice versa. In the biological situation, when two cells A and B are separated, if the cohesive strengths of their peripheries are about the same, then fragments of cell A will be left on the surface of cell B, and fragments of cell B will be left on cell A. This spatial difference between the adhesion interface and the plane of separation is illustrated on an extracted tooth, which is re-

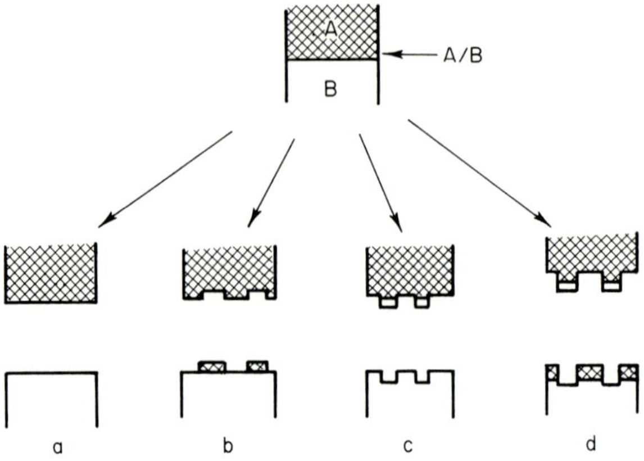

Fig. 1. The effect of the cohesive strength of two adherends A and B on the plane of separation, relative to their adhesive interface, A/B. (a) Cohesive strength of A equals that of B, and their peripheries are perfectly homogeneous. Separation occurs at adhesive (A/B) interface. Most improbable! (b) The cohesive strength of B is greater than that of A at points of true contact. Separation is accomplished by microruptures occurring in A, away from adhesive interface A/B. (c) The cohesive strength of A is greater than that of B at points of true contact. Separation is accomplished by microruptures occurring in B, away from adhesive interface A/B. (d) The cohesive strengths of A and B are about the same, and they adhere along an entire heterogeneous interface (A/B). Separation is accomplished by microruptures occurring in the peripheries of both adherends, leading to the plane of separation oscillating on both sides of the adhesive interface (A/B).

moved together with easily identified periodontal tissues, rather than cleanly, under normal circumstances.

A simple nontraumatic technique was developed to study the *in vitro* separation of cells from glass substrata (Weiss, 1961a). It was later demonstrated that when cells were detached from glass they apparently left behind material which, though invisible by light microscopy, could be detected by other cells (Weiss, 1961b). Later work by Weiss and Coombs (1963) and Weiss and Lachmann (1964) using immunological markers revealed residual material left behind on glass surfaces after cells were detached from them. Although fragments and/or tracks had been noted by other workers in relation to cell movement and detachment, their essential relationship to the mechanics of detachment was not fully appreciated.

It must be emphasized that one direct implication of this difference between adhesion and separation is that separation experiments do not tell us about the strength or nature of the adhesion between cells and their substrata; such experiments tell us about the separation process only.

Realization of this postulated difference between attachment and detachment, in that one is not simply the thermodynamic reverse of the other, allows a wider approach to contact problems than exclusive attention to adhesion. In ideal systems, adhesion can be analyzed in parameters of the cell periphery/environmental interface, such as surface charge, surface free energy (Steinberg, 1964), and critical surface tension (Baier *et al.*, 1968), whereas it is expected that separation will involve those factors affecting the cohesiveness of the peripheral zones of individual cells. A detailed analysis of the biophysical factors contributing to the cohesive strength or weakness of the cell periphery has been made recently (Weiss, 1967b) in connection with the separation of malignant cells from primary tumors. The main relevant point is that cohesive weakness, and hence cell separation, is favored by the development of structural heterogeneity as in the growth process, the activity of enzymes intrinsic or extrinsic to the involved cells, and the removal of calcium ions.

Various experiments in which cell separation from tissues is facilitated by treating them with calcium chelating agents are usually interpreted to mean that calcium is involved in cell adhesion in the form of radially oriented, divalent cationic "bridges" linking anionic sites on adjacent cells (Fig. 2A). Recently, it has been shown that cells which bind calcium to peripheral anionic sites, and which are, therefore, capable of being linked by calcium "bridges" in fact cannot be shown to exhibit calcium-sensitive adhesion. On the other hand, chelation of calcium from the peripheral regions of these cells is accompanied by facilitation of deformability, which indicates induced structural weakness. Thus, it has been proposed that a possible role of calcium, at least in short-term contact phenomena, is to raise the cohesive strength of the periphery of individual cells through tangentially oriented "bridges" (Fig. 2B). This rise in cohesive strength hinders the separation of one cell from another, as distinct from promoting the adhesion of one cell to another at the intercellular interface (Weiss, 1967d).

It is well known that enzymes facilitate the detachment of cells from each other and from noncellular substrata such as glass. Although such facilitation of detachment has in the past been interpreted exclusively in terms of enzymic activity at the site of interfacial adhesion between two adherends (e.g., Easty *et al.*, 1960), it appears that enzymic activity anywhere in the cell peripheral zone, by reducing cohesive

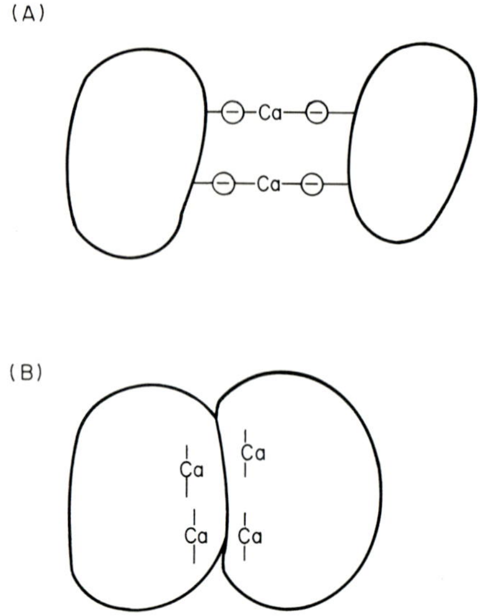

Fig. 2. (A) "Classic" radially oriented calcium bridges linking adjacent anionic groups (⊖) at the surfaces of two adjacent cells. As drawn, calcium ions will contribute to the strength of the adhesion between the two cells. (B) Tangentially oriented calcium bridges, linking sites with the peripheries of the same cells. As drawn, the calcium ions will increase the cohesive strength of the individual cells and, therefore, impede separation. These tangential bridges do *not* contribute to the strength of adhesion between the two cells.

strength, can facilitate cell separation (Weiss, 1963b). Support for this comes from the work of Weiss and Coombs (1963) in which by immunological techniques it was demonstrated that following prior incubation with trypsin, cells left species-specific antigen on the glass surfaces from which they were detached. These experiments indicate that in promoting cell detachment from glass, trypsin acts within the depth of the cell peripheral zone, as distinct from the cell/glass interface.

Following the work of Fell and Mellanby (1952) (in which it was shown that cultures of chick embryo limb-bone rudiments lose metachromatic staining matrix on exposure to excess vitamin A), Lucy, Dingle, and Fell (1961), demonstrated that the dissolution of cartilaginous matrix was due to the activity of enzymes having an optimal pH range of 3–5. Later, Dingle (1961) identified the enzymes as lysosomal hydrolases. It was then demonstrated that nonlethal activation of the lysosomal complex facilitated cell detachment (Weiss, 1965b). Weissmann (1965) has reviewed much of the literature on lysosomes and their activation during a very wide range of pathological processes. The concept of a chronic weeping lesion at cellular level, in which cells continuously promote their own detachment by an outpouring of lysosomal or other enzymes, appears particularly relevant to the situation in the periodontal sulcus, for example, where gingival cells are detached from the tooth elements.

The low optimal pH for lysosomal hydrolase activity has prompted the question of the value of pH near to the cell surface (Weiss, 1963c). Following the earlier work of Peters (1931), Danielli (1937) suggested that owing to charge effects, the hydrogen ion concentration near to a cell surface might well be lower than in the environment surrounding it. Hartley and Roe (1940) attempted to formulate the problem of "surface pH" more precisely and proposed that as the local concentration of ions around the surface of a particle is determined by its electrokinetic potential, ζ, the concentration of hydrogen ions in this region, relative to the concentration in the bulk phase (environment) is given by

$$[H^+_{surface}] = [H^+_{bulk}]e^{-\epsilon\zeta/kT} \tag{6}$$

At 30°C, Eq. (6) may be written as

$$pH_S = pH_B + \zeta/57$$

where pH_S = "surface" pH, and pH_B = bulk phase environmental pH. In the biological situation ζ has a negative value; thus, pH_S is lower than pH_B. Within the limitations discussed earlier, estimated values may be obtained for ζ, by measuring cellular electrophoretic mobility and transforming this into ζ by use of Eq. (5).

McLaren and Babcock (1959) have studied enzyme reactions in the presence of a number of negatively charged clay particles and have demonstrated significant differences between pH_S and pH_B by this use of enzymes as "molecular pH meters." In the case of cell surfaces where the possibility of electrokinetic heterogeneity exists, the unequivocal interpretation of studies of enzyme reactions in this region is presently impossible. The reaction between penicillin and the penicillinase of *Bacillus subtilis* in the cell-bound and free states was studied over a wide range of environmental pH (Weiss, 1963a). The results suggested that the bound penicillinase behaved as though it were reacting in an environment of lower pH value than that of the bulk phase. However, the observed effect was not as great as that calculated from measurements of the electrophoretic mobility of the organism. One possible way of interpreting the data is that many of the negatively charged groups in the peripheries of these cells are arranged in zones of higher than average surface charge density. An enzyme acting in the region between such high charge density zones would behave as though it were at a lower pH than in the bulk phase, but not in a region of as low pH as calculated from average electrokinetic data. On the other hand, an enzyme reacting with its substrate within a high charge density zone, would behave as though it were in an environment of lower pH than calculated from average electrokinetic data. It was suggested that the postulated zonal distribution of charged groups in the cell periphery may effectively localize and control its enzymic degradation. It must be emphasized that owing to technical limitations, this suggestion is highly speculative. While no estimates of the volumes of space involved in these considerations can be given, the distances from the surface charges over which these effects would operate are of the order of only a few Ångstrom units. It is, therefore, not expected that they could be detected with pH meter microprobe electrodes, or by common sampling procedures. In addition to these possible charge effects on local pH, the activity of other enzymes, from either the tissues or microorganisms, which generally tend to result in a fall of bulk phase pH in confined regions such as the periodontal sulcus, would also tend to maximize the effects of locally released lysosomal hydrolases, and other enzymes with low pH optima, on cell separation.

It is apparent from this brief review that although a general physical approach to cell contact phenomena can be formulated, a specific detailed examination of the phenomena is lacking. The main difficulty is that current experimental techniques give us estimates of variable reliability of *average* cellular parameters. These averages, which are hardly sufficient for the detailed, unequivocal analysis of "simple" nonliving systems, are grossly inadequate for the biophysical analysis of phenomena involving the heterogeneous, dynamic periphery of the living cell. However, the detailed recognition of difficulties may well portend advances in our knowledge of this fascinating field.

References

Ambrose, E. J. (1961). *Exptl. Cell Res.* Suppl. 8, 54-73.

Ambrose, E. J., ed. (1965). "Cell Electrophoresis." Churchill, London.

Baier, R. E., Shafrin, E. G., and Zisman, W. A. (1968). *Science* **162**, 1360-1368.

Bangham, A. D., and Pethica, B. A. (1960). *Proc. Roy. Phys. Soc. (Edinburgh)* **28**, 43-52.

Bikerman, J. J. (1961). "The Science of Adhesive Joints," 1st ed. Academic Press, New York.

Bishop, C., and Surgenor, D. M., eds. (1964). "The Red Blood Cell." Academic Press, New York.
Brooks, D. E., Millar, J. S., Seaman, G. V. F., and Vassar, P. S. (1967). *J. Cellular Physiol.* **69**, 155-168.
Curtis, A. S. G. (1962). *Biol. Rev.* **37**, 82-129.
Danielli, J. F. (1937). *Proc. Roy. Soc.* **B122**, 155-174.
Derjaguin, B. V., and Landau, L. (1941). *Acta Physicochim. URSS* **14**, 633-656.
Dingle, J. T. (1961). *Biochem. J.* **79**, 509-512.
Easty, G. C., Easty, D. M., and Ambrose, E. J. (1960). *Exptl. Cell Res.* **19**, 539-547.
Eisenberg, S., Ben-Or, S., and Doljanski, F. (1964). *Exptl. Cell Res.* **26**, 451-461.
Fell, H. B., and Mellanby, E. (1952). *J. Physiol.* (*London*) **116**, 320-349.
Hamaker, H. C. (1937). *Physica* **4**, 1058-1072.
Hartley, G. S., and Roe, J. W. (1940). *Trans. Faraday Soc.* **36**, 101-109.
Haydon, D. A. (1961). *Biochim. Biophys. Acta* **50**, 450-457.
Kauzmann, W. (1959). *Advan. Protein Chem.* **14**, 1-63.
Lisanti, V. F. (1960). *Ann. N.Y. Acad. Sci.* **85**, 461-466.
Lucy, J. A., Dingle, J. T., and Fell, H. B. (1961). *Biochem. J.* **79**, 500-508.
McLaren, A. D., and Babcock, K. L. (1959). *In* "Subcellular Particles" (T. Hayashi, ed.), pp. 23-36. Ronald Press, New York.
Mayhew, E. (1966). *J. Gen. Physiol.* **49**, 717-725.
Mayhew, E. (1967). *J. Cellular Physiol.* **69**, 305-310
Neiders, R., and Weiss, L. (1968). Unpublished.
Pauling, L. (1960). "The Nature of the Chemical Bond." Cornell Univ. Press, Ithaca, New York.
Perlitsh, M. J., and Glickman, J. (1966). *J. Periodontol.* **37**, 368-373.
Peters, R. A. (1931). *Proc. Roy. Soc.* **A133**, 140-154.
Pethica, B. A. (1961). *Exptl. Cell Res.* Suppl. 8, 123-140.
Pollack, W., Hager, H. J., Reckel, R., Toren, D. A., and Singher, H. O. (1965). *Transfusion* **5**, 89-95.
Pulvertaft, R. J. V., and Weiss, L. (1963). *J. Pathol. Bacteriol.* **85**, 473-479.
Rappaport, C. (1966). *Proc. Soc. Exptl. Biol. Med.* **121**, 1022-1028.
Rappaport, C., and Howze, G. B. (1966). *Proc. Soc. Exptl. Biol. Med.* **121**, 1010-1021.
Steinberg, M. S. (1964). *In* "Cellular Membranes in Development" (M. Locke, ed.), pp. 321-366. Academic Press, New York.
Valentine, R. C., and Allison, A. C. (1959). *Biochim. Biophys. Acta* **34**, 10-23.
Verwey, E. J., and Overbeek, J. T. G. (1948). "Theory of the Stability of Lyophobic Colloids." Elsevier, Amsterdam.
Wallach, D. F. H. (1967). *In* "The Specificity of Cell Surfaces" (B. D. Davis and L. Warren, eds.), pp. 129-163. Prentice-Hall, Englewood Cliffs, New Jersey.
Warren, L., Glick, M. C., and Nass, M. K. (1967). *In* "The Specificity of Cell Surfaces" (B. D. Davis and L. Warren, eds.), pp. 109-128. Prentice-Hall, Englewood Cliffs, New Jersey.
Weiss, L. (1961a). *Exptl. Cell Res.* Suppl. 8, 141-153.
Weiss, L. (1961b). *Exptl. Cell Res.* **25**, 504-517.
Weiss, L. (1963a). *J. Gen. Microbiol.* **32**, 331-340.
Weiss, L. (1963b). *Exptl. Cell Res.* **30**, 509-520.
Weiss, L. (1963c). *Biochem. Soc. Symp.* (*Cambridge, Engl.*) **22**, 32-50.
Weiss, L. (1965a). *J. Cell Biol.* **26**, 735-739.
Weiss, L. (1965b). *Exptl. Cell Res.* **37**, 540-551.
Weiss, L. (1966a). *J. Natl. Cancer Inst.* **36**, 837-847.

Weiss, L. (1966b). *J. Cell Biol.* **30**, 39-43.
Weiss, L. (1967a). "The Cell Periphery, Metastasis and Other Contact Phenomena." North-Holland Publ., Amsterdam.
Weiss, L. (1967b). *Proc. 7th Can. Cancer Conf.*, pp. 292-315. Pergamon Press, Toronto.
Weiss, L. (1967c). *J. Cell Biol.* **33**, 341-348.
Weiss, L. (1967d). *J. Cell Biol.* **35**, 347-356.
Weiss, L. (1968a). *Exptl. Cell Res.* **51**, 609-625.
Weiss, L. (1968b). *Exptl. Cell Res.* **53**, 603-614.
Weiss, L., and Coombs, R. R. A. (1963). *Exptl. Cell Res.* **30**, 331-338.
Weiss, L., and Kapes, D. L. (1966). *Exptl. Cell Res.* **41**, 601-608.
Weiss, L., and Lachmann, P. J. (1964). *Exptl. Cell Res.* **36**, 86-91.
Weiss, L., and Woodbridge, R. F. (1967). *Federation Proc.* **26**, 88-94.
Weiss, P. (1961). *Exptl. Cell Res.* Suppl. 8, 260-281.
Weissmann, G. (1965). *New Engl. J. Med.* **273**, 1084-1090 and 1143-1149.
Wilkins, D. J., Ottewill, R. H., and Bangham, A. D. (1962). *J. Theoret. Biol.* **2**, 165-191.

CHAPTER 2

SURFACE PROPERTIES INFLUENCING BIOLOGICAL ADHESION

Robert E. Baier

I. Introduction

The mode of formation, the structural features, and the chemical properties of biological interfaces are still largely unknown. The situation is particularly unsettled with respect to biological adhesion, which underlies the organization of all living tissue. Efforts to learn the details of adhesive behavior in biological environments have received new impetus recently from programs in fields as diverse as surgical adhesives, dental restoratives, prosthetic implants, extracorporeal circuits, and marine fouling. Cell contact phenomena have been extensively reviewed in the biomedical literature, and the physical bases of adhesive phenomena have been reported in detail by many authors in physical chemical journals. As an interdisciplinary effort, a recent attempt was made to carry current knowledge of adhesion over to bioadhesional problems (Baier *et al.*, 1968).

This chapter has three goals: (1) to review briefly some basic relations between the surface properties of materials and adhesion, and to outline the quantifying concept of surface "wettability" called "critical surface tension"; (2) to demonstrate the utility of these relations when considering the initial interactions at some surfaces of biological interest; and (3) to glean from the large body of biomedical literature some illustrations of the role surface properties play in determining adhesion in biological circumstances.

A. *Surface Forces Localized*

Langmuir (1916, 1925) many years ago demonstrated the extreme localization of surface forces, primarily by study of films of organic compounds on water, and referred to this concept as "the principle of independent surface action." From many investigations of the wetting of organic solids and of inorganic solids coated with organic films, it has been demonstrated that—with some understandable exceptions—this principle is usually true. The nature and packing of the exposed groups of atoms at any surface determine its wettability, a characteristic feature generally independent of the nature and arrangements of the underlying atoms and molecules (Shafrin and Zisman, 1954, 1960; Fox *et al.*, 1953; Zisman, 1957). Similarly, the outermost atomic constitution of a material must dictate its initial interactions with its environment.

The basic explanation is that the surface atoms in most solids and liquids generally attract each other by highly localized force fields, such as the London dispersion forces (London, 1937) which vary inversely as the sixth power of distance. The influence of dispersion-type force fields becomes unimportant at a distance of only a few atomic diameters, so there can be little contribution to the force of adhesion by atoms not in the surface layers. A rarely encountered exception occurs when the residual field of force is much less localized because either ions or large uncompensated dipoles are present in the surface of a material. An example is found in the unexpectedly strong wetting behavior of terminally fluorinated fatty acid monolayers (Shafrin and Zisman, 1957).

B. Physical Forces Sufficient as a Basis for Adhesion

Tabor (1951) illustrated that the work of adhesion for any system in which complete spreading occurs is simply twice the surface tension of the fluid phase. Assuming that the field of force causing adhesion vanishes in at most a few atomic diameters, a simple calculation shows that the tensile strength of the adhesive joint must be much greater than the tensile strength of common organic materials. Therefore, the joint must break by cohesive failure (in one of the adhering phases) rather than at the adherent-adhesive interface. It can be safely concluded, then, that when a fluid makes a zero contact angle with a rigid phase, the thermodynamic adhesive strength at the solid–liquid interface will always be much greater than the tensile or shear strength of the bulk phases; this has been demonstrated with organic materials many times (Henniker, 1949). With biological materials, L. Weiss (1961, 1962), in a series of experiments intended to estimate the strength of adhesion of cells during their distraction from various substrata, showed cohesive failure of the joint in one of the adherends rather than at the cell–substratum interface.

The role of chemical bonds between the adhering phases is often discussed, particularly when biological materials are involved. An investigation of the adhesion of polymers to wool (Feldtman and McPhee, 1964) showed that the energy involved in physical adsorption is more than adequate to produce adhesive forces greater than the cohesive strength of either organic adhesive or adherend. It is, therefore, unnecessary (but not unlikely) that specific chemical affinities exist for initial adhesion of materials, but the wetting and spreading must be sufficiently good to insure intimate molecular contact of the two phases if physical forces are to suffice.

II. Wettability Considerations

A. Liquids as Adhesives

When two solid surfaces are separated by a thin layer of liquid which wets them both well (i.e., the liquid has a zero or near zero contact angle), considerable adhesion results. Budgett (1911) found, for example, that two well-polished steel discs, less than one inch in radius but completely wet with paraffin oil, could only be pulled apart by applying a force of some 20 kilograms. This effect arises from the fact that the good wetting (expressible as low contact angle, approximating zero) by the liquid produces a concave meniscus at the liquid–air interface, with the pressure within the liquid being less than the pressure in the gas phase.

So, if the contact angle of a liquid with a more rigid surface is small, and particularly if the liquid is viscous, a prompt adhesive action is obtained. Such a thin fluid phase between two surfaces would join them together, with the joint having considerable strength in tension but only such resistance to shear as is determined by the viscosity of the liquid film. The permanence of the joint can be influenced by viscosity change of the fluid layer through any of a number of mechanisms such as permeation or loss of diffusible species, formation or elimination of cross-links, chemical bonding, or degradation at the interfaces.

For good adhesion, the contact angle of the fluid and rigid phases must be zero, or nearly so, in order that the greatest area of contact between the phases be generated by the free spreading of the fluid phase over the rigid surface and into any pores or capillaries. The contact angle must also remain small throughout subsequent changes in the joint so that localized stresses, which may cause the joint to fail (Griffith, 1920, Mylonas, 1955), will be minimal. Thus, attention to the problem of wetting and spreading of liquids on solids has considerable relevance to the understanding of adhesion and adhesives. Most biological surfaces can readily conform to the additional requirement for coplanarity and/or efficient mating of contours.

B. Contact Angle and Wettability

Most biomedical scientists underrate the importance of contact angle measurements, mainly because the variations in physical conditions affecting such measurements are not widely understood. Only in the past two decades have publications containing significant, reliable information on contact angles and their variation with the liquid and solid constitution been made available, and most of these derived from the research of Zisman's group at the Naval Research Laboratory (Zisman, 1961, 1963, 1964). It is useful to restate briefly the concepts of "wettability" and contact angle before attempting to discuss the affect of wettability on adhesion.

The most useful description of wettability is still based on Thomas Young's classic treatment (Young, 1805) of the contact angle (θ) between a liquid drop and a plane solid surface. When $\theta \neq 0$, the liquid is nonspreading, but not completely so: Every common liquid spreads on every common solid to some extent. When $\theta = 0$, the liquid is said to completely wet the solid, and it spreads spontaneously at a rate depending on its viscosity and the regularity of the solid surface. It has been found that for small drops (not gravitationally distorted), θ is independent of the volume of the liquid drop, and that temperature variations do not play a major role. The contact angle is, thus, a useful inverse measure of the wettability of a surface, in the sense that small contact angles are obtained with highly (or, easily) wettable surfaces; the cosine of θ is a direct measure, in the sense that, as the value of $\cos \theta$ increases toward unity, the wettability also increases.

C. *Critical Surface Tension of Wetting* (γ_c)

Systematic studies of the contact angles for a wide variety of pure liquids on organic surfaces (Fox *et al.*, 1953; Zisman, 1957; Shafrin and Zisman, 1960) and on inorganic surfaces (Hare and Zisman, 1955; Fox *et al.*, 1955) revealed many regularities in the wettability. An empirical rectilinear relation between the cosine of the contact angle ($\cos \theta$) and γ_{LV}, the liquid-vapor surface tension, of a series of liquids was established, and this in turn led to the useful concept of the critical surface tension of wetting (γ_c). The critical surface tension is defined as the intercept of the extrapolated straight line plot of $\cos \theta$ vs γ_{LV} with the $\cos \theta = 1$ axis, as illustrated in Figs. 1 and 2, to be discussed later. This intercept is precisely defined only for a homologous series of liquids, but it has been learned that the data for a large variety of nonhomologous liquids closely approximate a straight line or collect around it in a narrow rectilinear band.

On some surfaces, liquids of a given class will deviate from this band as do the H-bonding liquids on the polyamides of the nylon family (Ellison and Zisman, 1954; Baier and Zisman, 1967, 1968) and the halogenated liquids on polystyrene (Ellison and Zisman, 1954). This provides us with valuable additional information about the nature of the surface, such as its hydrogen-bonding potential or its reactivity towards special classes of materials. It has been found, in general, that the critical surface tension inferred from measurements with a variety of liquids is a very useful parameter *characteristic of the solid only*, and that it provides an empirical ranking of materials according to their relative surface energies. For the complex materials, inspection of the $\cos \theta$ vs γ_{LV} graph supplements the γ_c parameter with information on specific interactions.

The importance and optimum values of the thermodynamic work of adhesion, calculable from contact angle data and known surface tension, have been considered in detail elsewhere (Zisman, 1962) with the conclusion that the maximum work of adhesion generally increases in the same order as do the values of γ_c. For most organic substances the calculated work of adhesion varies only by a factor of two to three. It thus appears that the γ_c, since it is more readily obtainable, is at least as useful an index of surface wettability and adhesiveness as the maximum value of the work of adhesion.

The widespread occurrence of the rectilinear relationship between $\cos\theta$ and γ_{LV} in the now large body of experimental data makes it possible to use the γ_c values directly defined by these plots to compare the wettabilities of surfaces. Indeed, a "wettability spectrum" has been published which conveniently summarizes the relation between γ_c and surface constitution (Shafrin and Zisman, 1960). Changes in wettability are shown to be dependent on both the atomic constitution and the physical packing of atomic groupings in the outermost surface layer. For instance, the critical surface tension value decreases as the degree of hydrogenation of the surface is increased.

Similar correlations with γ_c exist for changing ratios of halogen-to-carbon in the surfaces. Of particular interest is the fact that an increasing ratio of chlorine atoms in the surface increases its wettability, while progressive fluorination decreases γ_c. It has become clear over the years that the γ_c values for surfaces of related chemical composition fall in a remarkably narrow band of the "wettability spectrum." The effectiveness of substitution of individual elements in increasing the wettability of organic surfaces is as follows:

Fluorine	←——→	Low surface energy
Hydrogen		
Chlorine		↑
Bromine		│
Iodine		↓
Oxygen		
Nitrogen	←——→	High surface energy

In continued study of surfaces of known composition and packing, the suggestion has developed that γ_c may be calculated for any surface whose molecular constitution, atomic arrangement, and surface density are known; some success has already been achieved (Bernett and Zisman, 1967).

Biological materials have critical surface tensions across a large range. Adsorbed layers of fatty acids have exterior surfaces primarily made up of closely packed methyl groups, and are, thus, of low wettability. The hydroxyl-rich polymers like starch (Ray *et al.*, 1958; Scholz *et al.*, 1958) have high γ_c values of 40–45 dynes/cm, in good agreement with that of 43 dynes/cm reported for the oxygen-rich surface of polyethylene terephthalate (Ellison and Zisman, 1954); wool has a γ_c approximating that of the simpler polyamide Nylon 66, with $\gamma_c = 46$ dynes/cm (Ellison and Zisman, 1954).

D. *Low-Energy and High-Energy Surfaces*

Solids may be grouped into two broad classes according to their specific surface free energies. Hard solids have surface energies ranging from 500 ergs/cm^2 to many thousand ergs/cm^2, the higher values corresponding to materials with high melting point and great hardness. In contrast, soft solids have specific surface free energies generally below 100 ergs/em^2, as do most liquids. Thus, it is to be expected that liquids will generally spread freely on solids of high surface energy but might show nonspreading behavior on low surface energy solids. For ease of reference, materials in these two classes may be called "high-energy surfaces" and "low-energy surfaces."

Results on the mechanism of spreading on high-energy surfaces can be generalized as follows: Every liquid will spread on smooth, clean high-energy surfaces at ordinary temperatures unless there has earlier formed on the solid a film which has critical surface tension less than the surface tension of the liquid. Because of the highly localized nature of the forces between contacting phases, a monolayer of adsorbed molecules is always sufficient to convert the high-energy surface to one with the same wettability properties as the low-energy surface having the same surface composition. Many compounds release polar contamination products (for instance, as they decompose or hydrolyze) able to form low-energy surfaces, and it has been shown that the nonspreading property can be produced in nearly all pure liquids by the presence of suitable adsorbable components (Bigelow *et al.*, 1946, 1947). Extremely pure liquids may also exhibit nonspreading behavior because of what has been termed the phenomenon of "autophobicity" (Hare and Zisman, 1955). When the molecules of a pure liquid adsorb to a high-energy surface in such a way that they form a film whose critical surface tension of wetting is less than the surface tension of the liquid itself, the liquid is unable to spread because of its own adsorbed and oriented monolayer. Polar liquids which are not autophobic are those whose surface tensions are less than the γ_c of their own monolayers; most nonpolar liquids do not exhibit autophobic properties.

E. The Influence of Water

Recent investigations of the equilibrium contact angle, θ—observed at high relative humidity, against the surface tension, γ_{LV}, of pure organic liquids on glass and metals (Shafrin and Zisman, 1967a; Bernett and Zisman, 1968, 1969)—resulted in the determination of a critical surface tension of only 25–30 dynes/cm. This low value indicated that an adsorbed water layer converts nominally high energy surfaces to ones which behave like surfaces of low energy toward nonhydrophilic liquids. When glasses or metals were equilibrated at low humidities, the data for the identical liquids showed good spreading to be the rule. Since the spreadability of organic liquids on high-energy surfaces is limited by the presence of one or more condensed monolayers of physically bound water, independent of the chemical composition of the underlying substrate (e.g., a variety of glasses and metals), there is a strong possibility that water adsorbed on other high-energy surfaces similarly changes the wettability. Indeed, Johnson and Dettre (1966) and Shafrin and Zisman (1967b) have demonstrated by different methods that the clean surface of bulk water is itself a low-energy surface: Shafrin and Zisman (1967b) reported a critical surface tension value of about 22 dynes/cm. Since the γ_c value for bulk water is less than that for adsorbed water at high humidities, it is apparent that the orientation effect of substrates on the water is influential in determining in which low surface energy arrangement it becomes constrained.

The applicability of these relations and experimental methods to complex biological materials holds considerable promise. As reviewed in later discussion, there is abundant evidence that biological adhesion is significantly dependent upon proteins adsorbed at the joint interfaces. We can then turn our attention to a summation of results from recent contact angle investigations on some protein analogs.

III. The Wetting Properties of Polyamides

A. Nylons

Although the absence of side structures is an important difference between polyamides and proteins, the polyamide series, commonly known as the nylons, has one essential feature in common with the proteins: a covalently linked backbone structure of carbon and amide groups. In 1954, Ellison and Zisman first demonstrated that nylon 66 (the most common industrial material) was wet differently by hydrogen-bonding liquids and nonhydrogen-bonding liquids. The interpretation of this result, supported by other wetting studies and by frictional measurements (Bowers *et al.*, 1953; Ellison and Zisman, 1954), was that the nylon surface included amide links which retained their hydrogen-bonding potentiality. Returning to the nylon study, in a more recent expanded series of experiments (Baier and Zisman, 1967, 1968, 1970), we have been able to learn of the influence of variable spacing of the amide links (by using nylons with differing lengths of hydrocarbon spacers) and to confirm the hydrogen-bonding capacity of the polyamide surfaces. As a selected example from a larger body of data, Fig. 1 shows the cos θ vs γ_{LV} plot for polyglycine (nylon 2).

For purposes of the present discussion, we shall note two features of these results. First, liquids which are capable of participating in hydrogen bonds—for example, water, glycerol, formamide, and thiodiglycol—interact more strongly with the nylon surfaces than do lower-surface-tension organic liquids. This is manifested in each case by smaller angles of contact of the H-bonding liquids than would be expected on the basis of results with the organic liquids alone: For non-amide polymers, contact angle values for both H-bonding and non-H-

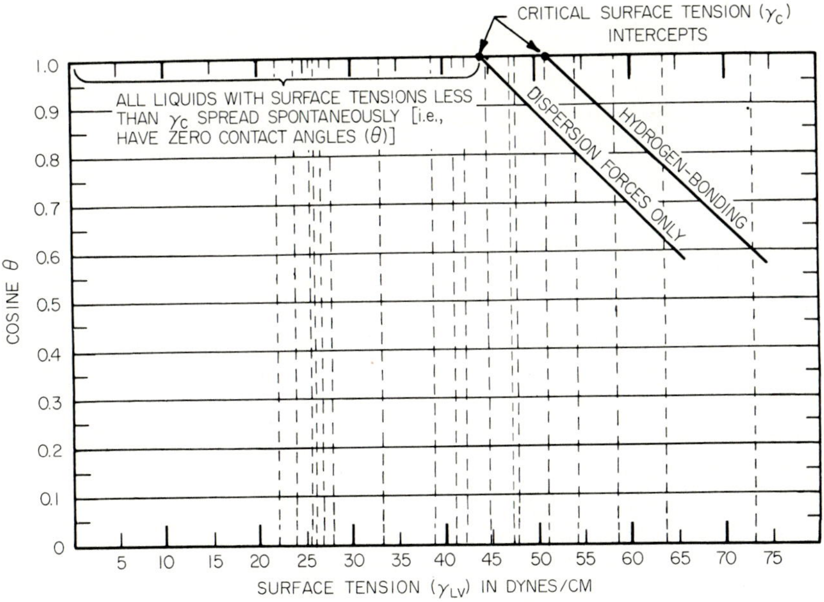

Fig. 1. Wettability of polyglycine film (nylon 2).

bonding liquids generally plot on the same straight line. The second feature is that the interaction of all of the liquids, regardless of their H-bonding character, is greater on the nylon 2 surface than on most other organic surfaces. This is manifested by smaller contact angles (or better spreading of the liquid over the solid), and reflects the fact that nylon 2 has the minimum of one methylene spacer separating amide links while other nylons, for example, have more. The conclusions from this set of observations are, first, that amide links are intrinsically more surface reactive than hydrocarbon groups and, second, that materials capable of hydrogen-bonding will interact even more strongly with polyamide substances because of such capability. Since nylon 2 (i.e., polyglycine) is the simplest model protein, we might presume that proteins in general are high surface energy materials like polyglycine. Certainly the H-bonding mechanism could assist in adhesion to these surfaces. Even in the presence of water the wettability of polyamide materials—including proteins—may be influenced by both the presence and the spacing of the amide groups in the backbone chain.

B. Polyacrylamide

In polyacrylamide, a water-soluble polymer whose amide groups occur in side chain substituents rather than in the polymer backbone, a strong interaction of these polar groups with water is found (Jarvis *et al.*, 1964). From the critical surface tension viewpoint developed in the previous sections of this report, it is clear that a very sharp distinction should be observed in the wetting of a polyacrylamide surface by H-bonding and non-H-bonding materials if adsorbed water does not overwhelm the surface interactions. From the accessibility and high relative density of amides in the surface, it is expected that the cos θ vs γ_{LV} plot would show separate grouping of the contact angle data for the H-bonding and non-H-bonding liquids. These data do plot on two separate lines, as in the case of the nylons, leading to different critical surface tensions (Jarvis *et al.*, 1964; Jarvis, 1967; Baier and Zisman, 1970).

A major new feature appears with polyacrylamide, however: Low surface tension organic liquids show small, but still measurable, contact angles on this surface in a region well below the nominal critical surface tension defined by either group (H-bonding or non-H-bonding) of liquids. This is completely anomalous when compared with the known polymer constitution. Past experience has shown that contact angle results are unusually reliable indicators of the nature of a plastic surface; so the polymer constitution, by itself, must not be indicative of all that is present in the polyacrylamide surface. When sources of organic contamination have been carefully excluded, attention must be focused on the high water affinity of this polymer. Data were obtained on polyacrylamide in a dry box held at less than 1% relative humidity, and on the same sample equilibrated at 99% RH (Jarvis *et al.*, 1964). The results may be interpreted as evidence that (1) even at less than 1% RH water is so strongly sorbed (and probably ordered) at the polyacrylamide surface that it gives rise to nonwettability by low-surface-tension organic liquids, and (2) even in the presence of high relative humidities, the hydrogen-bonding potentiality of the amide groups can be expressed.

We have thus learned a little more about the influence of water on some simple protein analogs. The limitations of these models are numerous, however. Generally, simple polyamides like the nylons, for instance, are extended chain structures which are intermolecularly hydrogen-bonded. What are the effects of intramolecular bonding? Nylons have no side chains to interfere with access to the polymer backbone; in polyacrylamide, the disposition of the side chains must be completely exterior, regardless of the conformation of the backbone.

The most serious inadequacies of simple polyamides as models for protein structure are apparent: Proteins have myriad side chains whose array is dictated by a variety of backbone configurations. Does the conformation of the backbone chain have an influence on wettability? Can the side chains mask the potential hydrogen-bonding contribution of the common polyamide backbone structure? As if to complicate our task, and make the eventual understanding of the surface properties of biological systems even more remote, both these questions have been answered in the affirmative by studies of a better protein analog, polymethylglutamate, which is a polypeptide with ester side chains.

C. *The Wetting Properties of Polymethylglutamate*

Poly-γ-methyl-L-glutamate (PMG), like many proteins, can assume a variety of conformations ranging from completely α-helical through a random-tangle structure to the extended chain β-configuration. In chloroform solution, for example, PMG is almost completely α-helical; films of the polypeptide dried from such solutions retain this secondary structure in the solid state (Loeb and Baier, 1967, 1968). Contact angle results for a typical α-helical film of PMG have shown that, despite the presence of the ester side chains, this polypeptide has wetting characteristics remarkably similar to those of the nylons (Baier and Zisman, 1967, 1968, 1970). It is especially interesting that the hydrogen-bonding capability is apparently retained. Molecular models of PMG in the α-helical structure illustrate the open access that the wetting liquids have to the polyamide backbone when this conformation is assumed. Molecular model studies also suggest that the random-tangle conformation allows accessibility to the polyamide backbone. It was, therefore, encouraging to find that PMG, when dried from a randomizing solvent, dichloroacetic acid, formed films with wettabilities nearly identical to those for the films cast from helix-inducing solvents (Baier and Zisman, 1967, 1968).

At the opposite structural extreme, however, scaled molecular models show that accessibility to the polyamide backbone of PMG is severely limited in the β-configuration. The α-helical form of PMG was conveniently transformed to the extended chain β-configuration by soaking the polypeptide films in formic acid. This conformation change was readily followed by infrared spectroscopy, and the newer technique of multiple attenuated internal reflection (MAIR) spectroscopy (Harrick, 1967) allows us to verify that the structural transformation was essentially complete at the immediate interfacial layer of surface chemical interest. In such transformed films, the wettability shows a remarkable change correlated with this modification of the polymer morphology. The anomalous (as compared with other polymers) interaction of the hydrogen-bonding liquids with this polyamide material disappears; the contact angles for all liquids on the transformed surface are greater. These results show that a hydrogen-bonding potentiality has been eliminated from the surface, and that certain reactive (or high surface energy) molecular groupings contribute less to the surface properties than before. The interpretation of these results is that the polyamide chain becomes masked from the surface by an orderly arrangement of the ester side chains (required in the β-configuration), thereby eliminating the hydrogen-bonding sites and, simultaneously, the high-energy contribution of the amide links. Indeed, the wettability of polymethylglutamate in the β-form is practically identical to that of the nonamide-based polymer, polymethylmethacrylate (Jarvis *et al.*, 1964), which has similar ester side chains and nearly the same hydrocarbon to ester ratio. This further illustrates the effective manner in which some protein side chains might shield more reactive functional groups from surface activity. It also points to a mechanism by which the surface controlled aspects of biologic adhesion might be either diminished or magnified as a result of (perhaps reversible) configuration changes in the proteins at cell or substrate interfaces.

D. Proteins

Deliberate structural modification of certain collagenous proteins, say by heat or chemical application, could be used to assist or impede adhesion of synthetic materials to many protein-rich structural tissues at the discretion of the operating surgeon. The high local concentration of water in biological materials will influence, but probably not overshadow, the wettability changes accompanying conformation changes. The demonstration of a conformation dependent wettability, although first achieved using a simple water insoluble polypeptide (Baier and Zisman, 1967, 1968, 1970) in which the influence of absorbed or adsorbed water is minimal, has been extended in a direct study of collagen and its heat-denatured product, gelatin (Baier *et al.*, 1968; Baier and Zisman, 1967, 1970). These studies confirmed, at least for the biologically important group of collagenous macromolecules, the conclusions reached from investigating the conformationally altered polymethylglutamate specimens: The wetting characteristics of proteins reflect changes in the molecular configuration even in the presence of strongly sorbed water.

It is important to stress the limitations of this experimental approach, however, especially in that an air phase has been present at all times. The major gain from contact angle work has not been an absolute determination of surface energy, but rather a *relative* ranking of surface properties of some materials of biological interest; only further work will tell if these results have value in predicting and controlling the behavior of such materials *in vivo*.

In spite of the lesser confidence one can have in experiments with natural materials, it appears that one *can* demonstrate differences in the surface properties of proteins with different structures. An intriguing observation referred to by L. Weiss (1959a) was that cells could be successfully cultured in contact with native serum, but not in contact with heat denatured serum. Perhaps a "critical surface tension" change accompanying configuration changes of one of the serum proteins may one day be correlated with this alteration in cell-substrate compatibility. This observation also suggests a mechanism by which naturally occurring, nonantigenic proteins could so readily change their surface properties during their adsorption at foreign surfaces that they now appear as foreign substances themselves. These now foreign proteins would activate the antibody and rejection systems which are responsible for the biological incompatibility of various implants. The antibody production against these now foreign proteins (adsorbed to nonphysiologic surfaces) is probably also manifested in interactions with nearby cells and tissues, and perhaps such a mechanism underlies the relatively minor—but effective—spermatozoa modifications which take place as these motile cells swim through the cervical zone modified by intrauterine contraceptive devices (Baier and Lippes, 1970).

There would be many advantages in knowing how to exert better control over surface interactions *in vivo*, either to promote or inhibit adhesion. We would like to improve adhesion of filling materials in teeth and of the apposed faces of cut tissue; we would like to impede adhesion of bandages to some wound surfaces and of marine biota to structural materials. The concept of tailor-making surface-active modifiers holds considerable promise for eventually meeting these needs (Zisman, 1965; O'Rear *et al.*, 1968; Bascom, 1968).

IV. Discussion

A. Recognition of the Role of Surface Forces in Biological Adhesion

1. Cell-to-Cell

Surface forces were recognized many years ago by Fahraeus (1929) as contributing to the mutual adhesion of red blood cells and to their aggregation into rouleaux. Fahraeus' microscopic inspection of the aggregation process provided observations supporting a physical mechanism tending to minimize the corpuscle-plasma interface, as opposed to a specific "stickiness" between the cell surfaces.

Surface interactions also seem to dominate the initial stage of contact among other types of cells. Initial contact between cells has been observed to be essentially random and nonspecific, with any cell potentially adherent to any other cell or foreign surface (L. Weiss, 1964). Numerous experiments on cell aggregation (Holtfreter, 1947; Lucey and Curtis, 1959) have demonstrated this randomization of initial contacts, but since the ultimate result in these studies is usually a sorting out of cells into specific tissuelike assemblages, it is apparent that the quality of adhesion between different cell types must be variable.

In cancer research, attention has been given to the possibilities that cell-to-cell interactions might be modified by surface changes promoting or inhibiting cell proliferation into tumor masses. Curtis (1962) has given a brief review of these phenomena, called "contact inhibition" and "contact promotion." L. Weiss (1962, 1967) has designed and carried out numerous experiments to test the possible relation of cell surface properties to the spread of cancer, particularly with regard to the diminished adhesiveness and increased invasiveness of malignant cells. There have been continued difficulties, however, in applying experimental data as tests of these concepts (L. Weiss, 1968a,b).

Physical scientists studying the surface-chemical phenomena of coagulation of colloidal dispersions (Hogg *et al.*, 1966) and the coalescence of immiscible liquid droplets (Torza and Mason, 1969) in model systems have also begun to point up the relevance of their measurements to biological contact phenomena.

2. CELL-TO-FOREIGN SURFACE

There is increasing evidence that the surface properties of materials, particularly with respect to their surface energy and wettability, are determinants of initial cell spreading and adhesion to these surfaces. P. Weiss (1958) noted, for instance, that one of the most important of cell activities, phagocytosis, could be viewed as fundamentally similar to cell spreading on a solid substrate; the spreading of the cells on a foreign surface was likened to "a vain attempt of the cell dwarf to engulf the substrate giant" by a mechanism similar to its engulfing smaller particles (P. Weiss, 1958).

Copley (1960) suggested that contact angle and surface tension changes were contributory factors in the interaction of blood elements with vessel walls, and recorded observations showing that modification of the surface properties by adsorbed films of fibrin and low-energy coatings were reflected in changes of the wall adherence and apparent viscosity of the blood. Although Leininger (1965) argued that thrombogenicity and blood clotting were not directly related to surface wettability, Lyman and co-workers (1965, 1968) have recently provided a correlation of blood coagulation times and platelet adhesion to polymers with the major parameter of the polymer wettability, their critical surface tensions.

In the work of L. Weiss (1960) and Taylor (1961, 1962), particularly, the influence of surface properties of polymers on the adhesion and spreading of cells in culture has been dealt with explicitly. In other cell cultivation work, the surface properties of polymers have been recognized implicitly (Tarasov and Bashmakov, 1966). Taylor (1961, 1962) has shown that cells attach and spread immediately on clean surfaces of glass, quartz, and some plastics. As will be discussed in a subsequent section, this attachment and spreading could be strongly inhibited when components from horse serum were adsorbed to the substratum. The speed with which cell attachment occurred, and the lack of observable specificity among a variety of cells and surfaces, suggests that physiological response of the living cells was not a major factor in initial cell adhesion. Indeed, in Taylor's studies (1961), living and dead cells showed similar contact behavior at both clean and serum-coated surfaces. On the other hand, in studies of L. Weiss (1964) in which dead cells did exhibit different contact relations from living ones, the cell surface charge did not vary significantly after cell death, thus diminishing the potential role of electrical properties in these initial contact phenomena.

So, a case can be made for the contention that the *initial* adhesion and spreading of cells is dictated by the particular surface properties existing at the cell-substratum interface, and, as observed, it should not be surprising that surface modifications, even by different samples of the same type of protein, have variable effects on cell spreading (Taylor, 1962). The common generalization from such observations, that cells adhere better to hydrophilic than to hydrophobic surfaces, would be better expressed by remarking that initial cell contact is correlated with rather simple considerations of surface wettability. L. Weiss and Blumenson (1967) discussed this approach, with special mention of the fact that attempts to relate cell adhesion to surface wettability (of the unmodified substrate) suffer from lack of knowledge of the *actual* surface properties of the substrate in the presence of the copius surface-active components of *in vivo* environs.

A further factor encouraging the comparison of biological adhesion to adhesion in nonliving systems is the observation that cell adhesion generally fails in cohesion—as do nonbiological joints. L. Weiss (1962) has shown, as mentioned earlier, that the separation of cells from substrata to which they have adhered is usually a cohesive failure. Other experiments have shown that the attachment of mollusk adductor muscles and ligaments to their calcareous shells is so firm that these joints also fail in either the organic or mineral phase, but not at the interface (Wakefield, 1964).

B. Examples of the Potential Correlation of Biological Interactions with Critical Surface Tension

Taylor (1961, 1962) observed the rate and degree of spreading of cells on a number of different surfaces. By listing the distributions of spread cells on each surface along with the water affinity of each surface, and showing no obvious correlation, he discounted the suggested relationship between surface wettability and cell contact behavior. The water affinity was ranked according to the water contact angle of each surface.

There are two deficiencies in that approach. First, the water contact angle is notoriously unreliable as an indicator of relative surface energy, particularly—in the writer's experience—for biological materials. Second, the water contact angles apparently obtained, as deduced from Taylor's order of water affinities, were at variance with other surface chemical measurements on the materials used. Actually, by viewing Taylor's cell spreading results (1961, 1962) in the context of the critical surface tension concept presented here, they provide an impressive demonstration of the probable correlation of surface properties with biological adhesion.

Figure 2 presents the Zisman plots of the contact angle data for five of the eight surfaces upon which cell spreading was observed: glass, cellulose acetate, adsorbed silicone, Vaseline, and Teflon. The inset table on Fig. 2 summarizes the cell spreading on each of these surfaces, derived from Taylor's graphs (1961, 1962), and the critical surface tensions, γ_c, derived from the intercepts of the Zisman plots at the $\cos \theta = 1$ axis. Although details of the derivation and preparation of Taylor's substrates were not presented, it is worthwhile to note briefly the origin of the surfaces which this writer used to produce the data resulting in Fig. 2, since they may differ from Taylor's in some important—but now unknown—way. The glass surface was that of a precleaned Bioloid glass microscope slide (Will Scientific Corp., Buffalo, New York) taken directly from a freshly opened package. The cellulose acetate was a thin transparent film, drop spread (Baier and Zisman, 1968) from a refined powder (Eastman Chemical Products, Kings Port, Tennessee) of the material in glacial acetic acid and thoroughly air flushed in a greasefree container. The adsorbed silicone was a coating of Siliclad (Clay-Adams, Inc., New York) baked onto a glass slide according to the manufacturers' instructions. The Vaseline was a thin, smooth film of this brandname product stroked over a glass slide with a warm spatula. The Teflon was a solid sample with a specularly smooth surface formed against glass optical flats.

Although Taylor (1961, 1962) ranked cellulose acetate as having the greatest water contact angle, that ranking is inconsistent with the result shown here. Similarly the ranking proved inconsistent with the critical surface tension intercept of about 34 dynes/cm, a result from studies with a variety of purified wetting liquids, including water, glycerol, formamide, thiodiglycol, methylene iodide, symtetrabrom-

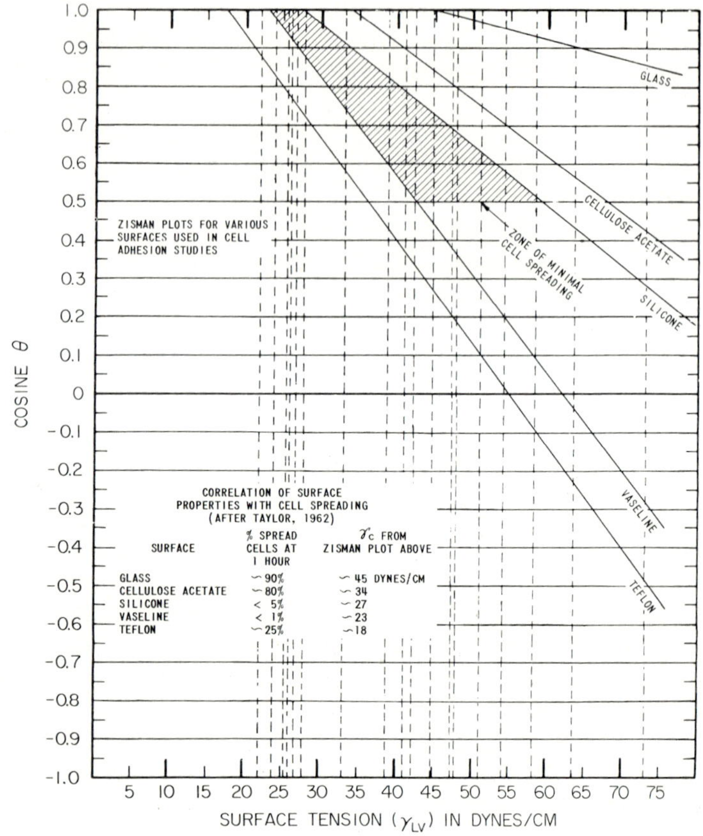

Fig. 2. Zisman plots of contact angle data for five surfaces upon which cell spreading was observed (see text).

oethane, l-bromonaphthalene, *O*-dibromobenzene, l-methylnaphthalene, dicyclohexyl, and hexadecane. It is also inconsistent with the γ_c range between 33 and 49 dynes/cm which has been reported for cellulose and its derivatives (Ray *et al.*, 1958; Scholz *et al.*, 1958; Shafrin, 1967; Luner and Sandell, 1969). Trace hydrophobic contaminants may have been responsible for Taylor's observation, or an anomalously large water contact angle may have resulted from the drop buildup phenomenon discussed in wetting studies on collagen and gelatin (Baier and Zisman, 1967) and since found with many water swellable polymers (Baier, 1969).

Another anomaly in the original ranking (Taylor, 1961, 1962), according to water contact angle, was the apparent smaller contact angle of water on Teflon compared with those on silicone and Vaseline. As the result of numerous careful observations (Fox and Zisman, 1950; Bernett and Zisman, 1959), it was found that water exhibits at least an equal, and usually larger, contact angle on Teflon than on any non-fluorinated surface. The origin of this anomaly is not known, but it also appeared in an independent report (Leininger *et al.*, 1966) where a photograph of water resting on "virgin" Teflon was published showing a contact angle in the range of 75°, when it should have been around 110°.

If then one chooses to ignore the questionable water-affinity ranking, and instead to pay attention to the typical critical surface tensions for the materials used, it is seen that Taylor's cell-spreading results can be usefully reinterpreted. The higher energy surfaces, glass and cellulose acetate, with their stronger liquid-solid interactions (as indicated by their high γ_c's and shallow slopes on the Zisman plot) induce the greatest and most rapid cell spreading. The low-energy surfaces, at least down to a critical surface tension of about 22 dynes/cm, correlate with minimal cell spreading. It is possible that the increased cell spreading on Teflon reflects a departure of the surface energy—this time toward the lower surface-energy extreme—from a probable optimum zone for minimal cell-substrate adhesion. Although it is admittedly speculative at this time, the liberty has been taken of indicating the probable zone of minimal cell spreading (and, by extension, maximum biological compatibility)—say, from 20 to 30 dynes/cm—on Fig. 2. This zone would include paraffin surfaces and medical grade silicones, for example. It might also include the critical surface tension value for the vascular endothelium (Baier *et al.*, 1969). Alternately, this biocompatible zone might extend downward to all critical surface tensions less than about 30 dynes/cm if the cell spreading studies on Teflon are not typical.

The other three surfaces studies by Taylor (1961, 1962) were adsorbed films of protein, which probably could not be readily reproduced in other laboratories and so were not included in Fig. 2. Nevertheless, it is worth noting that films of fibrinogen, which Taylor reported to show a high degree of cell spreading, exhibit a critical surface tension of about 36 dynes/cm (Baier and Dutton, 1969). This high degree of cell spreading agrees with the prediction which would have been made on the basis of its critical surface tension.

Other findings which may be explainable on the basis of critical surface tension have been reported. For example, hydrocarbon surfaces of polymers and adsorbed phospholipids, with their low γ_c values allow platelet adhesion but do not cause a severe enough interfacial reaction to rupture the platelets (Mustard *et al.*, 1967). On the other hand, when platelets come into contact with collagen, a protein with a γ_c of about 39 dynes/cm (Baier and Zisman, 1967, 1970), they not only adhere to it but become swollen and develop membranal flaws. L. Weiss and Blumenson (1967) have also demonstrated differences in cell adhesion to foreign surfaces of high and low surface energy, with adhesion and spreading being least on the low γ_c surfaces such as Teflon and siliconized glass, and greatest on clean glass. Finally, it may be recalled that marine larvae also exhibit considerable diversity in their adhesive behavior (Crisp and Ryland, 1960) which definitely appears to be associated with the critical surface properties of the substrates upon which they settle.

The contention will certainly be made that critical surface tension is, at best, an average property of the surface, and that the microheterogeneities on all known solids might be the most important determinants of cell adhesive phenomena. But the observation of cell behavior at isotropic surfaces contradicts this objection. Rosenberg (1963) successfully studied cells in culture at liquid–liquid interfaces between balanced salt solutions and organic liquids such as silicones and halogenated hydrocarbons. The cell behavior was markedly sensitive to the chemical composition of the organic phase, in some cases spreading uniformly and aggregating and in others remaining more separated and assuming spindle shapes. The isotropy of the liquid surfaces, eliminating the heterogeneous steps or stresses which might exist in the case of solids, coupled with the observation of differing cell behavior on liquids having differing atomic groupings, shows that cell–surface interactions do reflect differences in average composition of the surface. Further, close-packed lecithin molecules at liquid-liquid interfaces were able to convert these interfaces into ones of sufficiently different character such that the cells no longer adhered well or spread on them (Rosenberg, 1963).

C. *The Major Influence of Adsorbed Films*

1. "Conditioning" Proteinaceous Films

The extreme localization of surface forces and the case which has been made for their importance in bioadhesion in this article should make emphatically clear the importance of intervening films in determining biological adhesion. The modification of the simple concepts, considered in the previous sections, by intercellular materials, especially proteins, is enormous and complex.

In the studies of L. Weiss and Blumenson (1967), differences in the value of the cell adhesion parameter were practically voided when serum was present in the cell suspension. A major change occurred for adhesion to the glass surfaces which suggests the lowering of the glass surface energy by adsorption of protein from the serum. In addition, it was observed that the presence of protein in the cell-suspending medium roughly equalized the cell adhesion and spreading on surfaces as intrinsically different as Teflon and glass (L. Weiss and Blumenson, 1967). The culturing of cells on nominally high critical surface tension surfaces, like glass, is markedly sensitive to the cleaning and pretreatment of the surfaces and to the variabilities in the adsorption of proteins at these surfaces (Nordling *et al.*, 1965; Myllya *et al.*, 1966). Rappaport and colleagues (1960) hypothesized that the primary function of the serum in cell culture media was to modify the surface properties of the substrates, usually glass, by their adsorption at the solid-liquid interface.

The adsorption of blood proteins onto various solid surfaces has been shown to proceed almost instantaneously by a number of workers (Brash and Lyman, 1969; Baier and Dutton, 1969; Vroman and Adams, 1969; Scarborough *et al.*, 1969). The investigations of Lyman's group (Brash and Lyman, 1969; Lyman *et al.*, 1968) are particularly intriguing, because this same group demonstrated a correlation between the critical surface tensions of materials and the degree of platelet adhesion to them. If both results are valid, it means that the influence of substrate surface properties on cell adhesion is expressed *through* an adsorbed protein film of considerable thickness, certainly greater than a monolayer in depth (Baier, 1968; Baier and Dutton, 1969). The contention (Lyman *et al.*, 1968; Brash and Lyman, 1969) that the adsorbed protein films are undistorted monomolecular layers is not reasonable in light of the specific differences recorded for platelet adhesion to substrates which spontaneously adsorb proteins *before* platelet adhesion. [Platelet adhesion does not occur until about a minute after blood first contacts a foreign solid surface (Dutton *et al.*, 1968; Petschek *et al.*, 1968).] It is more likely that the adsorbed proteins are altered substantially, selectively, and sequentially during adsorption, in bucket brigade fashion (Vroman, 1964), and thereby present different outermost aspects (to other adsorbing proteins and to the settling platelets) which reflect the substrate-induced differences in their adsorbed configurations.

In other studies of platelet-substrate reactions, it was observed that protein adsorbed from plasma onto polystyrene modified platelet behavior from simple adherence to complex aggregation and rupture (Mustard *et al.*, 1967). Fibrinogen coatings on other proteins, including collagen, antigen-antibody complexes, and γ-globulin, seemed to modify the surface interactions enough to prevent platelet rupture but not platelet adhesion (Mustard *et al.*, 1967). With other cells, Taylor (1962) found γ-globulin to retard cell distortion (i.e., cell spreading) the most, and certain fibrinogen films the least.

Numerous other examples may be cited. It has been known for decades that blood proteins will adsorb strongly onto foreign surfaces and change their electrokinetic properties (Moyer and Gorin, 1940). This has been demonstrated again with candidate materials for surgical prostheses (Mirkovitch *et al.*, 1964). The initial modification of surfaces with heparin coatings (Gott *et al.*, 1963; Leininger *et al.*, 1966; Fourt *et al.*, 1966) has shown that this adsorption of protein can be reduced (Leininger *et al.*, 1966) but not eliminated. The initial adhesion of white cells and platelets to heparinized surfaces might also be reduced (Leininger *et al.*, 1966; Dutton *et al.*, 1968). The protein which has been most convincingly implicated in the initially adsorbed layers from blood is fibrinogen (Baier and Dutton, 1969; Vroman and Adams, 1969). It is presumably fibrin (i.e., polymerized fibrinogen) which makes up the bulk of the coating immediately and spontaneously formed on preserved homologous grafts and plastic prostheses in human and animal arteries (discussion following paper of Copley, 1960). Indeed, Dutton and co-workers (1969) have demonstrated that such an adsorbed "conditioning" film always precedes platelet deposition and the initiation of thrombus formation on foreign surfaces. Taylor (1967) gave independent evidence that adsorbed protein films from culture media intervene in the adhesion of other cells to foreign surfaces. Taylor's findings (1967), however, share a common difficulty with the recent results of Scarborough and co-workers (1969) in that Langmuir-Blodgett deposition (Langmuir and Blodgett, 1937) of extraneous protein films from air–water interfaces was not prevented. There have also been reports that blood platelets themselves are coated with an adsorbed protein which can modify clotting processes (Hjort *et al.*, 1955).

Lieberman and Ove (1958) demonstrated that a protein growth factor for a human cell line in tissue culture was required for the cells to adhere to glass surfaces and spread into a flattened shape. Presumably, the protein component adsorbed as a "conditioning" layer. L. Weiss (1959a) showed that serum was required for trypsinized cells to adhere to the surfaces of protein-free gels, but not to gels of the proteins fibrin and collagen; fibrin could also substitute for serum in conditioning agar and silica gels. Trypsinization of cells is an example of the effectiveness of enzymes in cell separation, probably through degradation of cell surface proteins. Thus, the requirement for added protein to reestablish cell adhesion is not unreasonable. Curtis (1962) reviewed other evidence for the involvement of proteins in biological adhesion, and cited work (Ambrose and Easty, 1960) interpreting enzyme action in cell separation as a manifestation of the influence of surface-adsorbed protein films.

Rosenberg (1960) demonstrated the deposition of a proteinaceous microexudate from cells which contact clean surfaces, and Taylor (1962) observed that cells, by means of such liberated products, inhibit the spreading and contact relations of themselves or nearby cells. Paul Weiss (1958) reviewed the involvement of extracellular intermediary layers in many cases of cell contact, and Rinaldini (1958) concluded that carbohydrate and protein macromolecules played the major role in intercellular layers in animal tissues.

It is the opinion of this writer that all these observations will eventually find a common explanation, based upon the induction of specific macromolecular configurations (and thus specific new outermost chemical arrays) determined during adsorption to various substrates. It is, unfortunately, not yet known what relationships exist between parameters of the substrate surfaces, like their critical surface tensions, and the configurations of absorbed biopolymers.

2. Surface-Active Adhesion Modifiers

Taylor (1962) noted that the presence of adsorbable materials in all body fluids, and the ability of cells to contribute additional organic materials to surfaces they contact, were strong inducements for study of the interaction of cells with surfaces of known properties. Other than this, the influence of adventitious surface-active species has gone unheralded in biology, although not unrecognized.

The thermodynamic tendency toward the minimization of a system's free energy is reflected in the reduction of the free surface energy of materials by diffusion of surface-active species to the interfaces from the adjacent bulk phases. A very small concentration of molecules having organophobic and organophilic groups suitably separated can accumulate as an adsorbed film at most organic interfaces, and the general features of such surface activity have been described in a number of publications (Ellison and Zisman, 1959; Jarvis and Zisman, 1959, 1960a,b).

The deliberate modification of foreign surfaces of some plastics currently used in biomedical applications, by incorporating diffusible surfactants in their bulk, has already met with success (Jarvis *et al.*, 1964). Thus, the way to the creation of poorly adhesive or more adhesive biomedical surfaces is open. In general, the surface-active materials naturally present at the site of a biological surface (to which adhesion should be aided or inhibited) will not have chemical constitutions compatible with obtaining the desired γ_c. Even the chemical constitution of conventional synthetic adhesion promoters does not meet the requirement for a high enough γ_c that common organic liquids will wet that surface completely and spread spontaneously; but it has recently been discovered that the incorporation of chlorine atoms in the outermost structure of adsorbed materials can significantly increase their γ_c (Shafrin and Zisman, 1968).

On the other hand, it is known that incorporation of fluorine atoms in the outermost surface of an organic material will significantly decrease its wettability. This is an example of "abhesion." "Abhesives" are films or coatings used to decrease the adhesion of one material to another making contact with it. Some of the common materials used industrially as "parting agents" are fluorocarbons, silicones, and long-chain fatty acids, alcohols, and amines. As little as a condensed monolayer of these materials will suffice to obtain the optimum effect. In view of the background information already presented here, it is evident that each of these films converts the surface to one of low surface energy, typically having γ_c values around 20 dynes/cm. Thus, adhesion is inhibited because most materials will no longer completely wet the surface. The poor surface wettability also allows the formation of voids or pockets of foreign materials at the interface, and the adhesion is further greatly reduced because of the local stress concentrations at the discontinuities. It is, therefore, evident that the best abhesives are those which have the largest possible contact angles for substances placed against them. It is probable that the adventitious formation of abhesive films in the mechanically prepared dental cavity is responsible for the poor adhesion and void entrapment by many new filler materials in modern dentistry.

The most likely surface-active adhesion modifiers for promoting strong bonds to hard tissues are polymers (Lee, 1967; Galligan and Schwartz, 1967). Eirich (1969) has recently reviewed and highlighted the special features of polymers as interface modifiers and is best quoted directly on this subject:

> . . . the multipoint adsorption of polymer molecules on sites up to 100 or more angstroms apart renders their attachment to the interface particularly insensitive to surface variations. Similarly, if the interface offered is not smooth or not quite compact, partial diffusion of chain molecules into most adsorbed substrates, with parts remaining protruding, allows chain polymer molecules to become very intimately tied, on a molecular scale, into any kind of surface structures.

There is another interesting aspect. The first dense polymer monolayer has no definite boundary against the next layer by virtue of the very variable entension of individual molecules away from the surface. Extended molecules reach into the next layer, or even into the third layer, and, vice versa, molecules from the "interior" of the melt will penetrate into the "adsorbed" layer. This continuous interweaving and intermeshing over several hundred angstroms' distance, between molecules which have segments truly absorbed and others which are merely entangled with adsorbed molecules, is unique with polymers and helps to build a substantial interphase.

D. *Involvement of Water*

A careful consideration of the various ideas about the structure of the cell surface suggests that the cell periphery consists of a zone of hydrated material. It is these hydrated zones which must interact when cells make mutual contact (L. Weiss, 1964, 1967). Fahraeus (1929) long ago proposed that adhesion between cells could be promoted by the increase of surface energy associated with desolvation of the cell surfaces by hydrophilic proteins. Schmitt (1941) followed this argument with evidence that the strong interaction between the basic protein histone, and the phospholipid, cephalin, could expel water from between the faces of cephalin bilayers and produce an insoluble complex. Addition of monolayer amounts of histone to red cell suspensions caused strong cell adhesion to develop (Schmitt, 1941). Schmitt (1941) recognized that, in general, all cell contacts would be influenced by the state of solvation of the cell surface. He especially noted the role which might be played by further desolvating protein interfaces, perhaps by calcium ions, in the promotion of cell adhesion. Although Steinberg (1962) argued effectively against simple dehydration of cell surfaces by calcium as an explanation for its known adhesion-promoting properties in some systems, the general proposition that Ca ions may interact specifically with hydrated protein surfaces, thus altering the molecular conformation and the hydrophilic-hydrophobic balance, still merits attention.

From the results presented in the foregoing, it can be inferred that it would be advantageous to use liquid biological adhesives, where applicable, which could either completely displace the water by principles of surface chemistry or react with or absorb the water. Many resins now in use as glass finishes to enhance resin to glass adhesion either react with or dissolve adsorbed water to some extent. From existing information on the water susceptibility of glass–resin reinforced plastic systems, however, it appears that the additional use of an appropriately designed and chemisorbed finish would be most advantageous (Shafrin and Zisman, 1968).

E. Other Complications

1. Theoretical Objections

Pethica (1961) provided a useful, critical discussion of physicochemical models for cell adhesion. He concluded that the treatment of cell adhesion as a wetting phenomenon (as developed in the 1960 review of L. Weiss) is probably not fruitful since the surface free energy of cell boundaries should be a function of curvature and location on the surface, and that contact angle relations are not reliable descriptions for the important localized complexes of cell membranes. As noted here, however, the γ_c concept does appear to allow useful empirical correlations.

Curtis (1960) adduced some rheological and adhesive properties which the cell surface may possess, and concluded that cells with like surfaces will form stable adhesion when separated by 100–200 Å *without the necessity for any intervening substance or intimate molecular contact.* His conclusions were based on considerations of the viscous and shear properties of cell surfaces, and extensions of the Verwey-Overbeek (1948) theory of the interaction of like surfaces. The specificity of adhesion and contact phenomena was controlled, in Curtis' scheme (1961), more by differences in time at which various cell types changed their surface properties ("temporal specificity") than by specific chemical arrangements at cell surfaces such as those proposed by Steinberg (1958).

Pethica objected to Curtis' conclusion (1960) that cell surfaces remain separated by 100 Å or more because calculations of Bangham and Pethica (1960) showed that cell surfaces can approach within 10 Å or less at portions of the periphery with small radii of curvature. He also objected to Curtis' estimate of the magnitude of the dispersion forces between cells. Curtis (1962) defended his arguments in favor of stable adhesions over large distances (ca. 100 Å) by challenging some of the assumptions in Pethica's calculations (1961). Curtis also argued that, since cells are likely to overlap most often with random orientation, the specific two-dimensional surface patterns proposed by Steinberg (1958) would lead to more variation in cellular aggregation that is actually observed.

It seems to this writer that these theoretical arguments are premature, and might better be postponed in favor of more carefully defining the interfacial conditions actually prevailing by experimental methods.

2. Auxiliary Role of Metabolic Processes

In a short discussion of cell adhesiveness, P. Weiss (1958) stated that wetting power if definitely involved, but not in any then discernible simple relation. He also stressed the active role of the cell in the expansion of its surface that accompanies adhesion. He further reasoned (P. Weiss, 1961) that, while the spreading of a cell at a solid-liquid interface is similar to the spreading of oil at an air-water interface, it is an oversimplification to consider the continued expansion of a cell surface in contact with a foreign surface to be a process dictated solely by laws of surface tension. He considered several lines of evidence which suggested that the cell expends metabolic energy in the course of its surface expansion, the net expansion of cells being controlled by competition between the cell cohesional forces and the adhesional forces (or "interfacial tensions") at the cell-substratum margin (P. Weiss, 1961; P. Weiss and Garber, 1952).

Moscona (1961) developed evidence that cell adhesion and cell-to-cell contact are temperature-dependent processes, reflecting the developmental state of the cells and the probable intervention of specific extracellular cell products (Moscona, 1962a). He also suggested (Moscona, 1962b), on the basis of low temperature studies and chemical inhibition of cell aggregation, that a metabolically produced adhesive substance underlay cell adhesion. Curtis (1963), on the other hand, showed that cells could reaggregate at temperatures as low as 1°C, where metabolic activity should be slight, if serum components were excluded from the culture medium. Later Curtis and Greaves (1965) isolated a pure serum protein which, unless destroyed by cell metabolism, could inhibit cell aggregation.

3. Drainage of Viscous Layers

If serum components act as a gap substance between cell and substrate, they could both prevent separation and hinder adhesion by virtue of their high viscosity which limits drainage from the interface region. This mechanism was invoked (Curtis, 1962) to explain the contradictory results of L. Weiss (1959a,b), who found that serum promoted strong cell-to-glass bonds not readily detached, and of Taylor (1961) who showed that serum could prevent the initial adhesion of cells to glass. In discussing the problem of drainage of extraneous viscous matter from between cell and substrate in adhesion, Curtis (1962) also noted that this process would be temperature dependent since the medium's viscosity would alter with temperature.

In the case of actively moving cells, however, the "drainage problem" of the viscous separating medium is minimized (L. Weiss, 1964). This is an example, nonetheless, of the possible confusion of cell adhesion and cell locomotion factors in the interpretation of observations. The association of adhesive phenomena with cell motility is one of the main difficulties in the definition of cell adhesion, and the two phenomena are nearly inseparable (Curtis, 1962).

4. Multivalent Ions

L. Weiss (1964) gathered evidence that cell-to-substrate joints do not fail at the interface, but in the weaker of the two adhering surfaces. This allows one interpretation of the role which calcium and other multivalent ions may have in bioadhesion: that of increasing the shear strength of cell surfaces by cross-linking sites on those surfaces (L. Weiss, 1964). The various ideas which have been proposed for calcium's action in bioadhesion were reviewed by Curtis (1962). Topological considerations applied to the binding of surfaces by calcium supported a nonspecific surface pattern.

Taylor (1962) quoted evidence that calcium ions are not required for the adhesion of some cells to glass (Easty *et al.*, 1960), and from his own work showed that the attachment of several types of normal stroma and epithelial cells would also occur in the absence of calcium. In an examination of the influence of calcium ions on spreading of cells, Taylor's experiments (1962) provided evidence that calcium was critical for cell spreading in the presence of serum proteins, although some variability in cell specificity was also noted. This supports the conjecture that the primary influence of multivalent ions on biological adhesion is through their modification of protein components at the interfaces.

V. Summary and Prospects

It has become apparent that interaction between cell and cell, or between cell and substrate, depends ultimately on intimate, perhaps molecular, contact between the interacting entities. So the need for an understanding of the constitution and reactivity of the contacting surfaces in bioadhesion, as in all adhesive systems, must be emphasized. In some cases, such as in the organization of tissue masses from individual cells, the interfacial properties contribute to strong mutual adhesion among certain cells but exclude others and simultaneously inhibit their spread beyond "normal" confines. In the "abnormal" case, typified perhaps by certain invasive and poorly adhesive cancer cells, it may be changes in the surface zone, possibly reversible, which eliminate the control often called "contact inhibition" which seems to be absent in metastasizing tumor masses.

In attempts to modify biological surface interactions, contradictory requirements will arise. We desire good tissue-to-plastic adhesion (or strong interaction) so that oral bandages will stay in place and some implanted prosthetic materials will be assimilated into the body structure. Simultaneously we desire surface properties that will minimize possible thrombus formation (through adhesion of platelets or activation of the clotting mechanism) and that will be compatible with the complex antibody and rejection mechanisms. In extracorporeal circulation we provide large surface areas to guarantee good gas and electrolyte exchange, and simultaneously ask that surface denaturation of crucial proteins be kept to a minimum despite the large surface contact between biological fluids and nonphysiologic materials. For surgical and dental adhesives, we require intimate molecular contact and good spreading to eliminate voids or particle entrapment while promoting strong adhesive bonds; for substitute heart valves we prefer to have blood-implant interactions kept to a minimum.

We should not presume that this challenge is too great. Similar contradictory requirements have arisen in fabric finishing technology, for instance, where soil repellency is desired to keep fabrics clean, but water compatibility is desired to encourage efficient laundering once soiled, although water repellency and soil repellency seem to be necessarily coupled. A new class of hybrid surface coatings has been discussed (Sherman *et al.*, 1969), however, which can change their configuration to present the most desired outermost chemical groupings for meeting these disparate environmental requirements.

There is the possibility that, in biological systems like blood vessels and joints, the ageing surfaces suffer just such a surface chemical alteration, but one which detrimentally increases γ_c, and so increases adhesion to the walls of materials circulating in the blood or present in other adjacent fluid phases (Baier *et al.*, 1968). We cannot evaluate this suggestion until more experimental measurements are made. New and more precise methods are needed for determining the surface tension of biological fluids and tissues in normal and pathological states (Perel'man, 1965). There are sharp differences, for instance, in the literature regarding the "wettability" of the endothelial lining of blood vessels, and its relation to the flow and coagulation properties of blood. Perel'man (1965) maintained that vessel endothelium is "nonwettable," in agreement with Moolten *et al.* (1949) and Ratnoff (1959), while Zubairov and co-workers (1963) and Copley (1962) rejected this thesis and its support for the idea that the liquid state of blood *in vivo* is controlled by the state of nonwettability of the endothelium. An important objection to the experimental methods of all of the above workers is that polar–nonpolar interfaces were produced in the blood, at which protein films and other surfactants accumulated in fractions of a second (thus modifying the interface); the "wettability" of the endothelial lining has only recently been determined by the critical surface tension criterion (Baier *et al.*, 1969).

In this enlightened age of molecular biology, it should be of some concern that we remain relatively ignorant of the structure and properties of molecules at biological interfaces. Without such knowledge, our attempts to meet the complex requirements of the above outlined problems are very limited.

It should now be clear how contact angle studies can provide one class of important information about the wettability and reactivity of materials of biological interest. A quantifying concept, called the critical surface tension (γ_c), has been introduced and its utility demonstrated by briefly reviewing the conclusions drawn over two decades of its study with well defined organic materials. The major features reflected in γ_c include the chemical composition of the surface and the presence of adsorbed species. By considering protein analogs such as the nylons and polypeptides, it was learned that the molecular configuration and the presence of potential hydrogen-bonding sites are also experimentally accessible surface properties. The study of collagen and gelatin cautiously extended these inferences to proteins. In the past few years, the role of water has been seen as contributing to a lower apparent critical surface tension (making the surface less wettable by nonhydrophilic liquids) for most materials. Therefore, fully hydrated biological interfaces seem to have general adhesive characteristics which reflect their poor wetting by most organic substances and their easy penetrability by H-bonding substances. It is emphasized that we still do not adequately understand the major influence which protein films have in biological adhesion. Continued study of proteins at interfaces seems to hold the greatest promise for immediate advances in our knowledge of this subject.

Acknowledgments

I am grateful to Dr. W. A. Zisman and his many co-workers at the Naval Research Laboratory, Washington, D.C., where I served a National Research Council Postdoctoral Associateship for supplementing my biophysics training with the numerous important surface-chemical concepts reviewed in the first portion of this chapter. I am also grateful to Dr. Leonard Weiss, of Roswell Park Memorial Institute, Buffalo, New York, for helpful discussions on cell adhesion and for suggesting many of the literature references discussed in the latter portion of this chapter.

References

Ambrose, E. J., and Easty, G. C. (1960). *Proc. Roy. Phys. Soc. Edinburgh* **28**, 53.
Baier, R. E. (1968). *Trans. Am. Soc. Artificial Internal Organs* **14**, (see discussion); following paper of Lyman *et al.* (1968).
Baier, R. E. (1969). Unpublished data.
Baier, R. E., and Dutton, R. C. (1969). *J. Biomed. Mater. Res.* **3**, 191.
Baier, R. E., and Lippes, J. (1970). In preparation.
Baier, R. E., and Zisman, W. A. (1967). *153rd Natl. Meeting Am. Chem. Soc., Miami Beach* p. 47 (Abstracts).
Baier, R. E., and Zisman, W. A. (1968). *Naval Res. Lab. Rept.* **6755**, 28 pp.
Baier, R. E., and Zisman, W. A. (1970). In preparation.
Baier, R. E., Shafrin, E. G., and Zisman, W. A. (1968). *Science* **162**, 1360.
Baier, R. E., Dutton, R. C., and Gott, V. L. (1969). *158th Natl. Meeting Am. Chem. Soc., New York* p. 100 (Coll. abstracts).
Bangham, A. D., and Pethica, B. A. (1960). *Proc. Roy. Phys. Soc. Edinburgh* **28**, 43.
Bascom, W. D. (1968). *Advan. Chem. Ser.* **87**, 38.
Bernett, M. K., and Zisman, W. A. (1959). *J. Phys. Chem.* **63**, 1241.
Bernett, M. K., and Zisman, W. A. (1967). *J. Phys. Chem.* **71**, 2075.
Bernett, M. K., and Zisman, W. A. (1968). *J. Colloid Interface Sci.* **28**, 243.
Bernett, M. K., and Zisman, W. A. (1969). *J. Colloid Interface Sci.* **29**, 413.
Bigelow, W. C., Pickett, D. L., and Zisman, W. A. (1946). *J. Colloid Sci.* **1**, 513.
Bigelow, W. C., Glass, E., and Zisman, W. A. (1947). *J. Colloid Sci.* **2**, 563.
Bowers, R. C., Clinton, W. C., and Zisman, W. A. (1953). *Lubrication Eng.* **9**, 204.
Brash, J. L., and Lyman, D. J. (1969). *J. Biomed. Mater. Res.* **3**, 175.
Budgett, H. M. (1911). *Proc. Roy. Soc.* **A86**, 25.
Copley, A. L. (1960). *In* "Flow Properties of Blood and Other Biological Systems" (A. L. Copley and G. Stainsby, eds.), pp. 97-117. Pergamon Press, Oxford.
Copley, A. L. (1962). *Proc. 8th Intern. Cong. Hematol., Tokyo, 1960,* Vol. 3, pp. 1648-1665. Pan-Pacific Press, Tokyo.
Crisp, D. J., and Ryland, J. S. (1960). *Nature* **185**, 119.
Curtis, A. S. G. (1960). *Am. Naturalist* **94**, 37.
Curtis, A. S. G. (1961). *Exptl. Cell Res.* Suppl. 8, 107-122.
Curtis, A. S. G. (1962). *Biol. Rev.* **37**, 82.
Curtis, A. S. G. (1963). *Nature* **200**, 1235.
Curtis, A. S. G., and Greaves, M. S. F. (1965). *J. Embryol. Exptl. Morphol.* **13**, 309.
Dutton, R. C., Baier, R. E., Dedrick, R. L., and Bowman, R. L. (1968). *Trans. Am. Soc. Artificial Internal Organs* **14**, 57.
Dutton, R. C., Webber, T. J., Johnson, S. A., and Baier, R. E. (1969). *J. Biomed. Mater. Res.* **3**, 13.
Easty, G. C., Easty, D. M., and Ambrose, S. J. (1960). *Exptl. Cell Res.* **19**, 539.
Eirich, F. R. (1969). *In* "Interface Conversion for Polymer Coatings" (P. Weiss and G. D. Cheever, eds.), pp. 350-373. American Elsevier, New York.
Ellison, A. H., and Zisman, W. A. (1954). *J. Phys. Chem.* **58**, 503.

Ellison, A. H., and Zisman, W. A. (1959). *J. Phys. Chem.* **63**, 1121.
Fahraeus, R. (1929). *Physiol. Rev.* **9**, 241.
Feldtman, H. D., and McPhee, J. R. (1964). *Textile Res. J.* **34**, 634.
Fourt, L., Schwartz, A. M., Quasius, A., and Bowman, R. L. (1966). *Trans. Am. Soc. Artificial Internal Organs* **12**, 155.
Fox, H. W., and Zisman, W. A. (1950). *J. Colloid Sci.* **5**, 514.
Fox, H. W., Hare, E. F., and Zisman, W. A. (1953). *J. Colloid Sci.* **8**, 194.
Fox, H. W., Hare, E. F., and Zisman, W. A. (1955). *J. Phys. Chem.* **59**, 1097.
Galligan, J. D., and Schwartz, A. M. (1967). *134th Ann. Meeting Am. Assoc. Advance. Sci., New York* p. 231.
Gott, V. L., Whiffen, J. D., and Dutton, R. C. (1963). *Science* **142**, 1297.
Griffith, A. A. (1920). *Phil. Trans. Roy. Soc. London* **A221**, 163.
Hare, E. F., and Zisman, W. A. (1955). *J. Phys. Chem.* **59**, 335.
Harrick, N. J. (1967). "Internal Reflection Spectroscopy." Wiley (Interscience), New York.
Henniker, J. C. (1949). *Rev. Mod. Phys.* **21**, 322.
Hjort, P., Rapaport, I., and Owren, P. A. (1955). *Blood* **10**, 1139.
Hogg, R., Healy, T. W., and Fuerstenau, D. W. (1966). *Trans. Faraday Soc.* **62**, 1638.
Holtfreter, J. (1947). *J. Morphol.* **80**, 25.
Jarvis, N. L. (1967). Private communication.
Jarvis, N. L., and Zisman, W. A. (1959). *J. Phys. Chem.* **63**, 727.
Jarvis, N. L., and Zisman, W. A. (1960a). *J. Phys. Chem.* **64**, 150.
Jarvis, N. L., and Zisman, W. A. (1960b). *J. Phys. Chem.* **64**, 157.
Jarvis, N. L., Fox, R. B., and Zisman, W. A. (1964). *Advan. Chem. Ser.* **43**, 317.
Johnson, R. E., Jr., and Dettre, R. H. (1966). *J. Colloid Interface Sci.* **21**, 610.
Langmuir, I. (1916). *J. Am. Chem. Soc.* **38**, 2286.
Langmuir, I. (1925). *In* "Third Colloid Symposium Monograph," pp. 48–75. Chem. Catalogue Co., New York.
Langmuir, I., and Blodgett, K. B. (1937). *Phys. Rev.* **51**, 317.
Lee, H. L. (1967). *134th Ann. Meeting Am. Assoc. Advance. Sci., New York* p. 231.
Leininger, R. I. (1965). *In* "Biophysical Mechanisms in Vascular Homeostasis and Intravascular Thrombosis" (P. N. Sawyer, ed.), p. 288. Appleton, New York.
Leininger, R. I. Epstein, M. M., Falb, R. D., and Grode, G. A. (1966). *Trans. Am. Soc. Artificial Internal Organs* **12**, 151.
Lieberman, I., and Ove, P. (1958). *J. Biol. Chem.* **233**, 637.
Loeb, G. I., and Baier, R. E. (1967). *Naval Res. Lab. Rept.* **6554**, 11 pp.
Loeb, G. I., and Baier, R. E. (1968). *J. Colloid Interface Sci.* **27**, 38.
London, F. (1937). *Trans. Faraday Soc.* **33**, 8.
Lucey, E. C. A., and Curtis, A. S. G. (1959). *Med. Biol. Illus.* **9**, 86.
Luner, P., and Sandell, M. (1969). *157th Natl. Meeting Am. Chem. Soc., Minneapolis* p. 11 (Cell. abstracts).
Lyman, D. J., Muir, W. M., and Lee, I. J. (1965). *Trans. Am. Soc. Artificial Internal Organs* **11**, 301.
Lyman, D. J., Brash, J. L., Chaikin, S. W., Klein, K. G., and Carini, M. (1968). *Trans. Am. Soc. Artificial Internal Organs* **14**, 250.
Mirkovitch, V., Beck, R. E., Andrus, P. G., and Leininger, R. I. (1964). *J. Surg. Res.* **4**, 395.
Moolten, S. E., Vroman, L., Vroman, G. M. S., and Goodman, B. (1949). *A.M.A. Arch. Internal Med.* **84**, 667.
Moscona, A. A. (1961). *Nature* **190**, 408.

Moscona, A. A. (1962a). *In* "Biological Interactions in Normal and Neoplastic Growth" (M. J. Brennan and W. L. Simpson, eds.), pp. 113-126. Little, Brown, Boston, Massachusetts.

Moscona, A. A. (1962b). *J. Cellular Comp. Physiol.* **60**, Suppl. 1, 65.

Moyer, L. S., and Gorin, M. H. (1940). *J. Biol. Chem.* **133**, 605.

Mustard, J. F., Glynn, M. F., Nishizawa, E. E., and Packham, M. A. (1967). *Federation Proc.* **26**, 106.

Myllya, G., Hayry, P., Pettinen, K., and Saxen, E. (1966). *Ann. Med. Exptl. Biol. Fenniae (Helsinki)* **44**, 171.

Mylonas, C. (1955). *Exptl. Stress Anal.* **12**, 129.

Nordling, S., Pettinen, K., and Saxen, E. (1965). *Exptl. Cell Res.* **37**, 161.

O'Rear, J. G., Sniegowski, P., and James, F. L. (1968). *Advan. Chem. Ser.* **87**, 10.

Perel'man, I. B. (1965). *Colloid J. (USSR) English Transl.* **27**, 422.

Pethica, B. A. (1961). *Exptl. Cell Res.* Suppl. 8, 123-140.

Petschek, H. E., Adamis, D., and Kantrowitz, A. (1968). "An Experimental Preparation for the Study of Thrombosis on Artificial Surfaces Under Controlled Flow Conditions." Avco Everett Res. Lab. Rept. No. 314.

Rappaport, C., Poole, J. P., and Rappaport, H. P. (1960). *Exptl. Cell Res.* **20**, 465.

Ratnoff, O. (1959). *In* "Connective Tissue, Thrombosis, and Atherosclerosis" (I. Page, ed.) Academic Press, New York.

Ray, B. R., Anderson, J. R., and Scholz, J. J. (1958). *J. Phys. Chem.* **62**, 1220.

Rinaldini, L. M. J. (1958). *Intern. Rev. Cytol.* **7**, 587.

Rosenberg, M. D. (1960). *Biophys. J.* **1**, 137.

Rosenberg, M. D. (1963). *In* "Cellular Control Mechanisms and Cancer," (P. Emmelot and O. Muhlebock, eds.), pp 146-164. Elsevier, Amsterdam.

Scarborough, D. E., Mason, R. G., Dalldorf, F. G., and Brinkhous, K. M. (1969). *Lab. Invest.* **20**, 164.

Schmitt, F. O. (1941). *Growth* **5**, Suppl., 1-20.

Scholz, J. J., Ray, B. R., and Anderson, J. R. (1958). *J. Phys. Chem.* **62**, 1227.

Shafrin, E. G. (1967). *In* "Polymer Handbook" (J. Brandrup and E. H. Immergut, eds.), pp. 111-113. Wiley (Interscience), New York.

Shafrin, E. G., and Zisman, W. A. (1954). *In* "Monomolecular Layers," (H. Sobotka, ed.) p. 129, Am. Assoc. Advance. Sci., Washington, D.C.

Shafrin, E. G., and Zisman, W. A. (1957). *J. Phys. Chem.* **61**, 1046.

Shafrin, E. G., and Zisman, W. A. (1960). *J. Phys. Chem.* **64**, 519.

Shafrin, E. G., and Zisman, W. A. (1967a). *J. Am. Ceram. Soc.* **50**, 478.

Shafrin, E. G., and Zisman, W. A. (1967b). *J. Phys. Chem.* **71**, 1309.

Shafrin, E. G., and Zisman, W. A. (1968). *Advan. Chem. Ser.* **87**, 20.

Sherman, P. O., Smith, S., and Johannessen, B. (1969). *157th Natl. Meeting Am. Chem. Soc., Minneapolis* p. 13 (Cell. abstracts).

Steinberg, M. S. (1958). *Am. Naturalist* **92**, 65.

Steinberg, M. S. (1962). *In* "Biological Interactions in Normal and Neoplastic Growth" (M. J. Brennan and W. L. Simpson, eds.), pp. 127-140. Little, Brown, Boston, Massachusetts.

Tabor, D. (1951). *Soc. Chem. Ind. (London), Rept. Progr. Appl. Chem.* **36**, 621.

Tarasov, V. N., and Bashmakov, G. A. (1966). *Vopr. Virusol.* **11**, 492.

Taylor, A. C. (1961). *Exptl. Cell Res.* Suppl. 8, 154.

Taylor, A. C. (1962). *In* "Biological Interactions in Normal and Neoplastic Growth" (M. J. Brennan and W. L. Simpson, eds.), pp. 169-182. Little, Brown, Boston, Massachusetts.

Taylor, A. C. (1967). *134th Ann. Meeting Am. Assoc. Advance. Sci., New York* p. 230.
Torza, S., and Mason, S. G. (1969). *Science* **163**, 813.
Verwey, E. J. W., and Overbeek, J. T. G. (1948). "Theory of the Stability of Lyophobic Colloids." Elsevier, Amsterdam.
Vroman, L. (1964). *Thromb. Diath. Haemorrhag.* **10**, 455.
Vroman, L., and Adams, A. L. (1969). *J. Biomed. Mater. Res.* **3**, 43.
Wakefield, H. F. (1964). *Proc. Symp. Exper. Marine Ecol.* No. 2, 51.
Weiss, L. (1959a). *Exptl. Cell Res.* **17**, 499.
Weiss, L. (1959b). *Exptl. Cell Res.* **17**, 508.
Weiss, L. (1960). *Intern. Rev. Cytol.* **9**, 187.
Weiss, L. (1961). *Exptl. Cell Res.* Suppl. 8, 141.
Weiss, L. (1962). *J. Theoret. Biol.* **2**, 236.
Weiss, L. (1964). *J. Theoret. Biol.* **6**, 275.
Weiss, L. (1967). "The Cell Periphery, Metastasis, and other Contact Phenomena." North-Holland Publ., Amsterdam.
Weiss, L. (1968a). *Exptl. Cell Res.* **51**, 609.
Weiss, L. (1968b). *Exptl. Cell Res.* **53**, 603.
Weiss, L., and Blumenson, L. E. (1967). *J. Cellular Physiol.* **70**, 23.
Weiss, P., (1958). *Intern. Rev. Cytol.* **7**, 391.
Weiss, P. (1961). *Exptl. Cell Res.* Suppl. 8, 260-281.
Weiss, P., and Garber, B. (1952). *Proc. Natl. Acad. Sci. U.S.* **38**, 264.
Young, T. (1805). *Phil. Trans. Roy. Soc. London* **65**, 65.
Zisman, W. A. (1957). *In* "A Decade of Basic and Applied Science in the Navy," p. 30. U.S. Govt. Printing Office, Washington, D.C.
Zisman, W. A. (1961). *Naval Res. Lab. Rept.* **5699**, 28 pp.
Zisman, W. A. (1962). *In* "Symposium on Adhesion and Cohesion" (P. Weiss, ed.), p. 176. Elsevier, Amsterdam.
Zisman, W. A. (1963). *Ind. Eng. Chem.* **55**, 19.
Zisman, W. A. (1964). *Advan. Chem. Ser.* **43**, 1-51.
Zisman, W. A. (1965). *Ind. Eng. Chem.* **57**, 26.
Zubairov, D. M., Repeikov, A. V., and Timberbaev, V. N. (1963). *Fiziol. Zhu. SSSR* **49**, 85.

PART II

BIOLOGICAL ADHESIVES

CHAPTER 3

ADHESION OF CELLS TO SURFACES

A. Cecil Taylor

I. Introduction

The cohesive unity of animal tissues is the result of two conditions. One of these is the existence of an intercellular matrix of mucinous, fibrous, or membranous elements which binds or ensheaths isolated and aggregated cells. The elaboration of this matrix is the province of the connective tissue cells. The second condition insuring structural unity is the surface adhesiveness of many cells themselves. This feature is most highly developed in the epithelial cells where the surface membranes have become specialized for attachment to other cells of their kind or, less frequently, to noncellular surfaces.

Examples of the latter are not numerous in higher animals. In this respect the area of contact between gingival epithelium and tooth enamel is of particular interest because this is the only place in the mammalian body where active cells of the germinative layer of an epithelium are so intimately related to an essentially inorganic solid. This close relationship comes about in the course of tooth development by the regression and loss of the layer of enamel-forming cells and its replacement by a thin keratinous cuticle, one which, incidently, may easily be removed by abrasion or by clinical manipulation.

Questions raised by this relationship are, "How firmly can cells adhere to such a noncellular surface, and what is the nature of the adhesion?" For reasons mentioned below, it has proved extremely difficult to determine accurately, in living man or experimental animal, the strength of the adhesion between the soft gingival tissue and the hard tooth which penetrates it. This chapter will consider the possibility of studying this relationship *in vitro* and the application of some of the observations made in an attempt to do this.

II. Relationship of Gingival Epithelium to Teeth

A. *The Gingival "Attachment Epithelium"*

The concept of a firm organic union between the tooth and adjacent epithelial cells was presented by Gottlieb (1921) early in this century and has for many years been generally accepted and taught in text books on oral histology. It was based upon the appearance in histological preparations of an intimate relationship between the enamel of the tooth crown and cells of the contiguous gingival epithelium which form a collar around the cervical area and have been designated the "attachment epithelium." Although it was stated by Orban (1949) that the cells "are firmly fastened to the tooth and must follow its movements," there has been no clear understanding of the character of such an attachment. In general some sort of cementing material (Sicher, 1962) or of fibrous bonding has been assumed. One explanation offered by Baume (1953) and based upon phase contrast studies was that tonofibrils of the involved epithelial cells extend into and become continuous with the enamel cuticle.

There has been considerable difference of opinion regarding the nature and even the existence of such an epithelial attachment. Primary among the reasons for this is the fact that it has been difficult to demonstrate clinically in man that such a strong attachment actually exists. It has also been difficult to reconcile such a firm attachment of its cells with the concept of the crevicular epithelium as a dynamic tissue in which cells proliferate in the basal layer, migrate toward the surface, and become desquamated into the gingival sulcus. That this does occur has recently been clearly demonstrated by the autoradiographic studies of Gruelich (1962) and of Skougaard and Beagrie (1962). Furthermore, studies with the electron microscope have shown that tonofibrils of epithelial cells do not cross their cell membranes (Odland, 1958; Kelly, 1966). They also provide no evidence of a substantial fiber linkage extending between epithelial cells and the enamel.

B. The Gingival "Cuff"

In the light of this type of disagreement and based upon extensive observations of his own, Waerhaug (1952) discredited the whole concept of the epithelial attachment and he now, along with others, contends that the cells are held only very weakly against the surface of the teeth by what he prefers to call adhesion. The term "epithelial cuff" was proposed as a more appropriate designation for the cells which adhere to the crown.

An apparent reason for the sharp difference in these opposed views was the lack of an objective criterion for measuring the strength of adhesion of living cells to the tooth. In recognition of this lack several attempts have been made to determine the force necessary to separate the attached cells from the tooth crown. Waerhaug (1952) inserted a thin steel blade into the sulcus and determined that a pressure of 7 gm was needed to carry the tip to the cemento-enamel junction. The same test was done by Parfitt (1961) using more sophisticated sensing equipment which registered a somewhat greater pressure required for the same penetration. In neither experiment, however, was it possible to be certain that the blade actually passed between the cells and the enamel surface. In fact when a cellulose acetate strip was inserted into the same position by Zander (1956) and left in place while the tooth was fixed and sectioned, later microscopic examination showed that the strip had been forced in between the cells of the epithelial cuff, leaving a layer of cells still attached to the tooth. These tests show the futility of efforts to make meaningful measurements of forces acting at the cellular or subcellular level on clinical material where microscopic visualization is nearly impossible.

III. Cell Adhesion

It is important to make clear what is implied by use of the term adhesion with respect to cells on a surface. The forces operating in cell adhesion have been investigated and evaluated by L. Weiss (1960), Pethica (1961), Curtis (1962), and others, and a summary of the biophysical aspects of cell contact has been presented in the chapter by Weiss, this volume.

It should be recognized that there are two phases of the phenomenon of cell adhesion. One of these is an expression of the physical attraction forces operating directly between the molecules of the cell membrane and those of the substratum surface, while the other results primarily from chemical bonds established between the surfaces or between each surface and molecules of substance which comes between them.

The intensity of the physical attraction forces is determined by factors such as (1) the molecular configuration and physical properties of the adhering surfaces, (2) the distance of their separation, and (3) their electrochemical properties. These adhesive forces and their governing factors are operative at the time of cell contact, and the resulting effect may be described as immediate adhesion. In the sense that the vital processes of the cells are not involved in the relationship, it is a passive attraction. Investigative observations on cell adhesion restricted to this early phase shortly after initial cell contact may detect only the passive physical aspects of cell-to-surface relationships.

Over a longer time interval the immediate adhesion may be altered by physiological changes in the living cells determined by their genetic makeup and expressed in chemical and structural modifications of the surfaces. Such alterations can either reduce adhesive strength or intensify it and make it more durable. The latter is thus a long-term adhesion in which the cells may play an important active role. Because of their firm bonding, cells held by long-term adhesion are often called attached cells, even though the attachment remains one of intermolecular forces. In the use of this term by cell biologists there is no implication of the existence of an organic union, or continuity with the material of the substratum such as was proposed for the "epithelial attachment" of the gingiva.

It should be pointed out that most investigations of cell contact and adhesion have employed either connective tissue cells which, even in their natural environment, lack any special adaptations for attachment to their surroundings, or cells that have been maintained in culture conditions for a prolonged time during which they could have lost their original capacity to form special adhesive mechanisms. In either case, the cells, while subject to normal immediate adhesion, might be incapable of developing a stronger attachment to a solid surface.

IV. Measurement of Adhesion by Detachment of Cells

The possibility of a different approach to the problem of cell adhesion to teeth was suggested earlier, that of bringing the epithelium and the teeth together in a tissue culture environment. Since under these conditions microscopic examination of the cells is feasible, measurements of their adhesion might be more accurately carried out. This approach would also provide an opportunity to compare the adhesive reaction of a variety of cell types to the tooth surface as well as to observe the response of gingival cells to surfaces other than that of the tooth. It would bring to the problem of relationship between gingiva and teeth information that has already been gained through many studies on cell contact which have been carried out on cells *in vitro*.

Various methods have been used to estimate or measure the strength of adhesion of cells in culture. All of these have attempted to evaluate adhesion by determining the force required to separate cells from a surface to which they had already become adherent. Cells have been detached by simple mechanical traction with a glass microneedle (Coman, 1961), by a stream of fluid through a tube (Dan, 1947), by gravity (Taylor, 1961), and by centrifugal force (Easty *et al.*, 1960). The most extensive and thorough studies, however, have been those carried out by L. Weiss (1961) who has used the shearing force generated by a metal disc immersed in the culture medium and lowered to a measured distance from the cells and then rotated at a known velocity.

A. *The Micro-Jet Shearing Method*

It would be difficult to apply any of these methods to the measurement of adhesion of cells cultured on the irregular surface of an opaque rat tooth. Furthermore, as we shall see, there are reasons why it would be desirable to compare the adhesion of different types of cells which emigrate from a single explanted fragment of oral mucosa. For these reasons in our investigation it was decided to detach cells individually while they were being observed under the microscope. This was done by lowering a micropipette into the medium and positioning it with respect to a selected cell so that a stream of culture medium, when forced through the pipette, would flow over the cell. The spatial relationships between the micropipette, jet stream, and cell are shown in the diagram of Fig. 1. The velocity of flow in this jet stream was increased at a constant rate from zero at the beginning of each test up to the moment of cell detachment when the final velocity was recorded.

Certain limitations of the method and problems in the interpretation of jet stream detachment have to be taken into consideration. Those which apply to the shearing component of the force exerted by the stream have been discussed by L. Weiss (1961). Other problems arise because of the involvement of lesser component forces, such as a horizontal thrust against protruding portions of isolated cells, a downward vertical thrust, and a possible Venturi lift. These would complicate greatly any attempt to compute the actual force exerted upon a particular cell. In view of these problems, plus the unknown effects of cell deformation on the physical properties of cytoplasm and the unpredictable changes in geometry of the cell and the impinging jet stream, there was little purpose in attempting to calculate absolute values for the force. However, with adequate attention to details such as angle of inclination of the pipette and distance of its tip from the margin of the cell, it was assumed that most of the unknown quantities would cancel out, and that significant comparisons could be made between cells, particularly when the minimal force required for their detachment differed consistently by an appreciable amount. For these comparisons, therefore, the velocity of the fluid as it emerges from the pipette has been used as an expression of the force required for detachment. It should be pointed out that the actual shearing force exerted upon cells would be proportional to the square of any change in velocity of the jet stream.

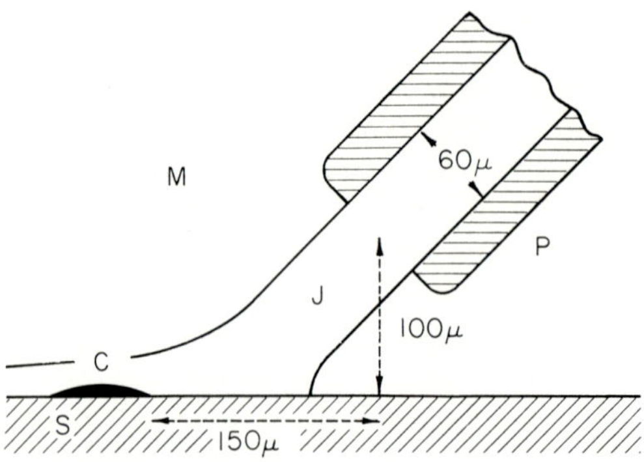

Fig. 1. Diagram of the geometry of the jet stream in relation to a cell. J = jet stream; M = liquid culture medium; P = tip of pipette; S = substratum; and C = cell.

The cells used for comparison of adhesive strength were obtained from stock strains of several different marmoset tissues currently being carried in our laboratory as well as from outgrowths of basal cells from the epithelium of oral mucosa taken from laboratory rats. The test cells were allowed to adhere to and spread on a variety of surfaces including glass, polystyrene, epoxy resin, and reconstituted rat tail collagen, as well as on the enamel surface of extracted rat molars and incisors. The primary cultures were tested after periods of incubation from three days to three weeks.

B. Results

The results of large numbers of measurements made over several years' time have shown that there is very little consistent difference in the strength of adhesion by any one type of cell to the different substances. Such a lack of difference in long-term adhesive response by cells to a series of surfaces physically and chemically so different from each other as those of this list has also been observed by L. Weiss (1962) in a similar experiment. This fact is particularly significant in view of the marked effect of even relatively slight differences in surface properties of the substratum upon immediate or passive adhesion (Rappaport and Bishop, 1960; Easty *et al.*, 1960; Taylor, 1961). It can be interpreted to indicate that the plane of separation passes through a substance existing between the cell membrane and the substratum, a substance which is similar in all tests regardless of the material of the substratum. Such a substance would presumably be produced by the cells themselves although it might also contain components obtained from the medium.

This extracellular material should be considered a product of, but not a part of, the cell since it occurs outside of the plasma membrane and may, in large part, be removed as by tryptic action without affecting the integrity or viability of the cell. Because of this covering, the cell membrane itself may normally not contact and adhere directly to any surface, attachment being mediated by the peripheral adhesive material. According to this concept a cleavage through this material at any level between the molecules of the cell membrane and those of the substratum surface could be considered a break in cell adhesion. When the strength of adhesion to the substratum exceeds the cohesive forces within the cytoplasm, the cell may rupture and leave part of its surface attached to the substratum. In such cases the adhesion, obviously, cannot be measured by the method of shearing.

Although cells exhibited similar adhesion to a variety of surfaces there was a marked difference in the adhesive attachment of different types of cells to the same surface. In general, cells from the established strains which were tested, regardless of their origin from connective tissue or epithelium, were all detached by jets with essentially similar rates of flow. As seen in Fig. 2, these cells were slightly less resistant to the shearing stress than fibroblasts that had recently emerged from fragments of explanted oral mucosa. The latter connective tissue cells, although healthy in appearance and well spread on the test surface, were removed by a jet velocity of less than 3 meters per second.

Most resistant to jet stream shearing were epithelial cells which had emigrated from explants of rat floor of the mouth, gingiva, or palate (Fig. 2). On the average these withstood a jet flow of 3–5 times the velocity which would detach connective tissue cells appearing in the same culture. Many withstood velocities as high as 15 meters per second.

Cells cultured on a transparent substratum could be viewed with transmitted light by bright field or phase contrast illumination (Fig. 3). However, in order to visualize cells spread on the surface of an

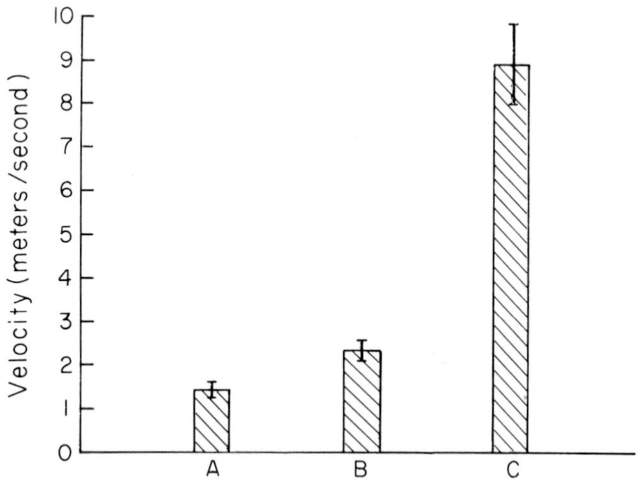

Fig. 2. Bar graph showing the mean velocities of jet stream required to detach the three types of cells. Central line represents the standard deviation.

	meters/second
A. Permanent strain cells	1.4
B. Primary culture connective tissue cells	2.3
C. Primary culture epithelial cells	9.0

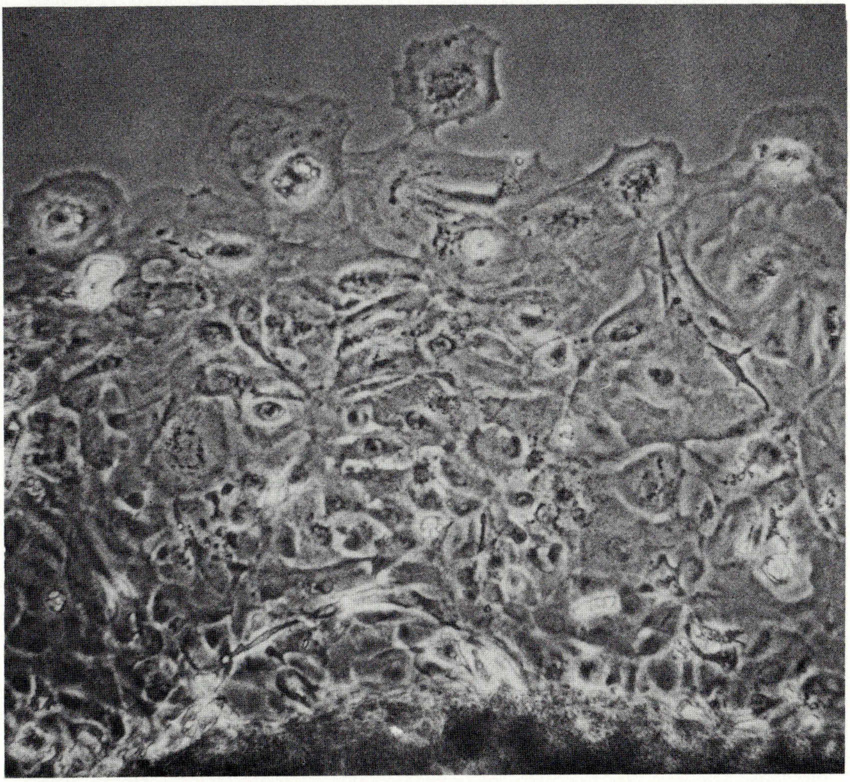

Fig. 3. Primary epithelial cells emigrating from the margin of a fragment of oral mucosa from a rat. ×300.

opaque tooth crown it was necessary to stain the cells first with a vital dye and view them by incident light. Methylene blue applied for a few minutes served to darken the cells sufficiently without affecting their viability or adhesive qualities. The appearance of cells adhering to rat molar enamel is shown in Fig. 4. In these outgrowths from tissue onto the tooth surface there was observed the same marked difference between the epithelial and connective tissue cells in respect to their adhesive tenacity. The presence or absence of a cuticle on the enamel appeared to make little difference in cell adhesion.

C. Cell Deformation

Certain features of cell deformation that were seen during detachment gave information about the character of adhesion. In general, connective tissue cells and established strain cells were not greatly

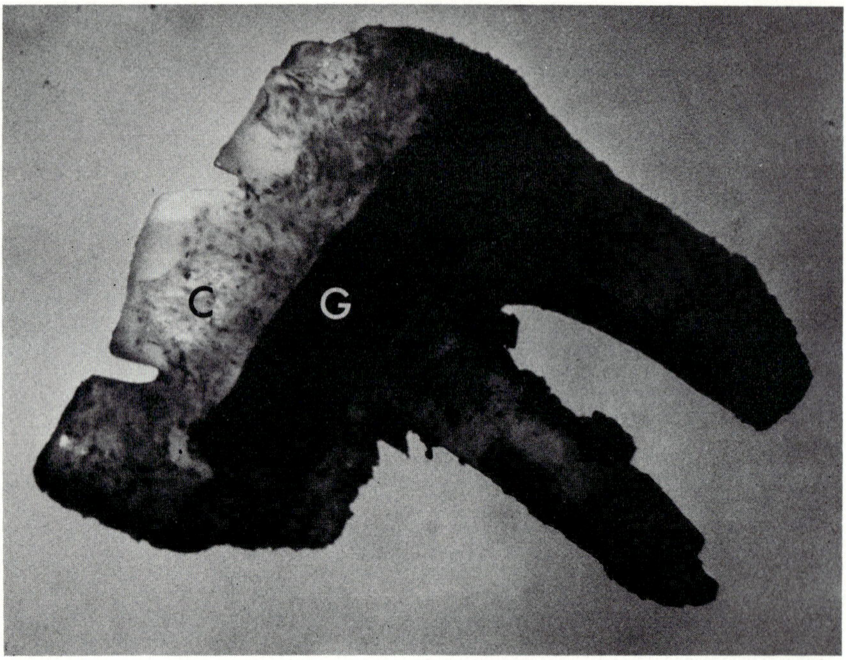

Fig. 4. Rat molar with stained cell outgrowth (C), from gingival fragment (G) cultured for 6 days. ×30.

distorted by the jet stream and remained attached until the flow of the stream reached a critical velocity, at which point they lost their attachment suddenly and were swept away. Larger and more flattened cells, particularly those in the epitheloid strain cultures, occasionally detached first along one margin after which the freed portion folded back. The remainder of the cell then peeled off from the substratum.

Connective tissue and established strain cells were not visibly ruptured or seriously injured by the process of detachment by shearing, an observation corroborated by the reattachment and spreading of such cells during a subsequent period of observation. On the other hand, epithelial cells from primary cultures of oral mucosa were affected differently. Their margins generally remained firmly attached until the pressure of the shearing stream was great enough to cause marked distortion of the cell. When the jet stream velocity had reached 4 meters per second, the nucleus and much of the cytoplasm was displaced. By 6 meters per second small portions of the cytoplasm were sometimes observed being detached from the cells. At stream velocities higher than this, portions of the cell margin began to detach from the substratum and peeling or tearing generally followed. Occasionally a cell remained adherent to the surface by only a small marginal attachment which would withstand pressures beyond the calibrated range of our equipment. Epithelial cells, detached singly or in sheets at stream velocities exceeding 6 meters per second, appeared moribund and would not spread thereafter. Frequently portions of marginal cytoplasm were left attached to the substratum after the major part of the cell had been swept away.

These jet stream detachment experiments have demonstrated that, under the conditions of these tests, the same amount of force was required to remove a particular type of cell from any of a variety of different surfaces, including the surface of a tooth. That is to say, the material of the substratum had little effect on the strength of adhesion. On the other hand, epithelial cells which had recently emerged from their tissue environment were capable of affixing themselves to a surface much more firmly than any of the other cell types.

V. Fine Structure of Adhering Surfaces

It was of considerable interest, therefore, to examine these cells with the electron microscope in the hope of determining whether any special features in the contact area could be found which correlated with the difference in their adhesive behavior and which might therefore furnish some clues regarding the mechanism of the adhesive phenomenon.

A. *Subcellular Space and Contact Areas*

Study of profile pictures of the cells adherent to epoxy resin showed that often only a small portion of the cell surface apposed to the substratum is in adhesive contact (Figs. 5, 6, 7). This agrees with the direct observations on living fibroblasts made by Ambrose (1961) using the surface contact microscope. Therefore, we closely examined those portions of the cell membrane probably in adhesive relationship as indicated by their close approximation to, and essentially parallel alignment with the substratum surface. Measurements made within the contact areas of each type of cell showed that the surface membranes of connective tissue and strain cells generally approached more closely to the resin surface than those of the epithelial cells. For the former, the characteristic distance was between 20 and 150 Å, whereas with primary epithelial cells the range was from 100 to 500 Å.

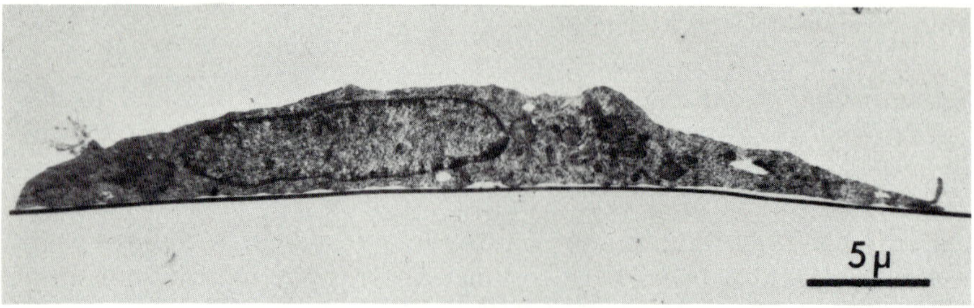

Fig. 5. Connective tissue cell emigrated from fragment of rat gingiva cultured 5 days on epoxy resin substratum. Note limited regions of actual contact with substratum. ×2880.

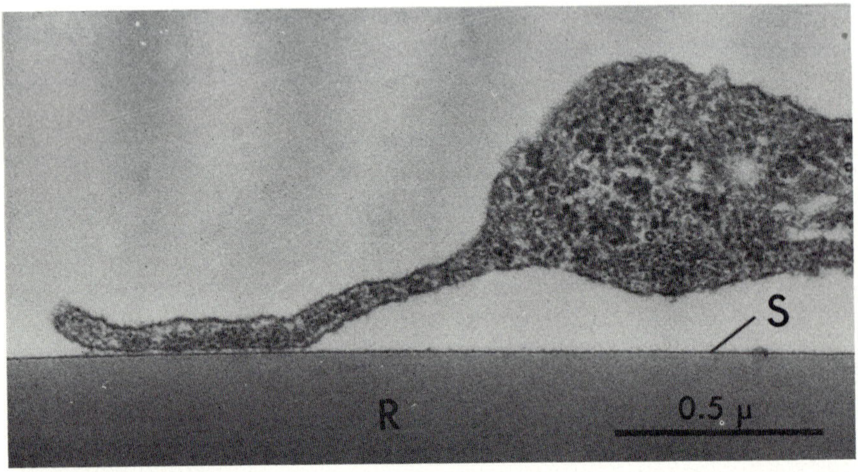

Fig. 6. Process of a connective tissue cell from rat gingiva cultured on epoxy resin. S = substratum film. Note adhesive area with extracellular material in space measuring 50-100 Å wide.

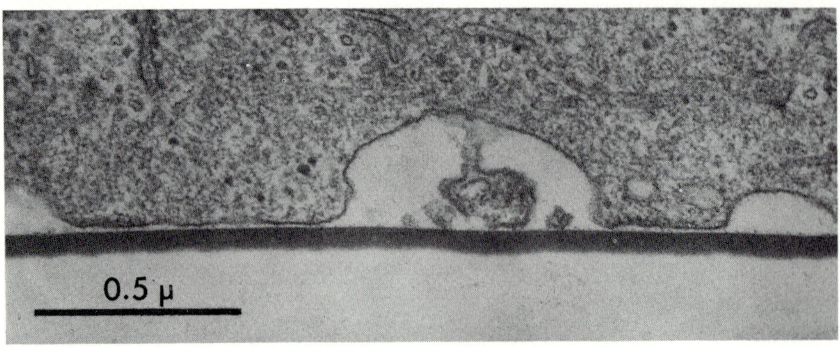

Fig. 7. Contact areas between washed marmoset liver strain cell and epoxy substratum cultured for 18 hours in serum free medium. Note lack of substratum film and small amount of extracellular material. ×57,000.

These measurements indicate that the stronger bonds of long-term adhesion cannot be attributed to physical forces of attraction acting directly between the unit membrane of the cell and the substratum surface. This is true because the distance over which such forces can act is limited to a very few Ångstroms (Curtis, 1962), but more obviously because the space between the surfaces is more or less densely occupied by electron defracting material.

B. Extracellular Material

The presence of an extraneous coat on the surface of all cells, possibly in part elaborated by the cells themselves and playing a role in their surface adhesion, has been generally recognized (P. Weiss, 1958; Rinaldini, 1959; L. Weiss, 1960; Moscona, 1961; Bennett, 1963; Rambourg *et al.*, 1966). Efforts to identify this substance have indicated that it contains mucopolysaccharides, proteins, and possibly phospholipids. It has been shown that a material produced by the cell may also be transferred to the surface of the substratum (Rosenberg, 1960) and that this can strongly affect the adhesive properties of the cells themselves (Taylor, 1961; L. Weiss, 1961; Rubin, 1966).

One component of the material present between the cell membrane and the epoxy resin surface in the culture experiments is an electron dense film having a thickness of about 50 Å and separated from the resin by an electron lucent space (Figs. 6, 8, 10). The free surface of

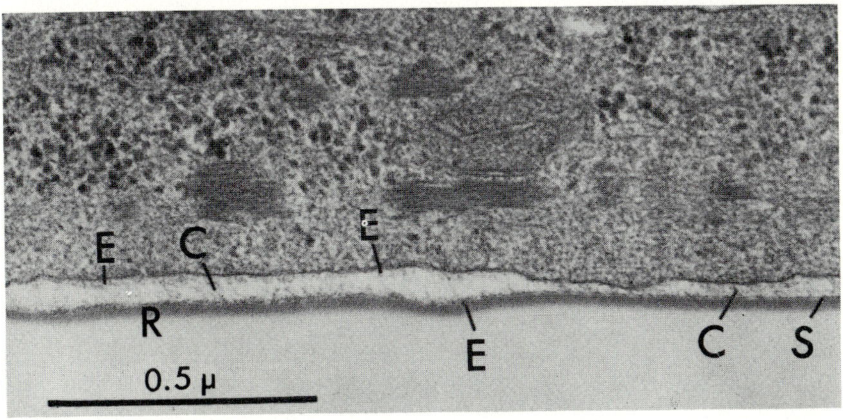

Fig. 8. Surface of epithelial cell from rat gingiva cultured on epoxy resin surface. R = osmium stained portion of resin substratum; S = substratum film; E = tufts of extracellular material adhering to cell and substratum surfaces; and C = strands connecting cell membrane with substratum film. ×72,500.

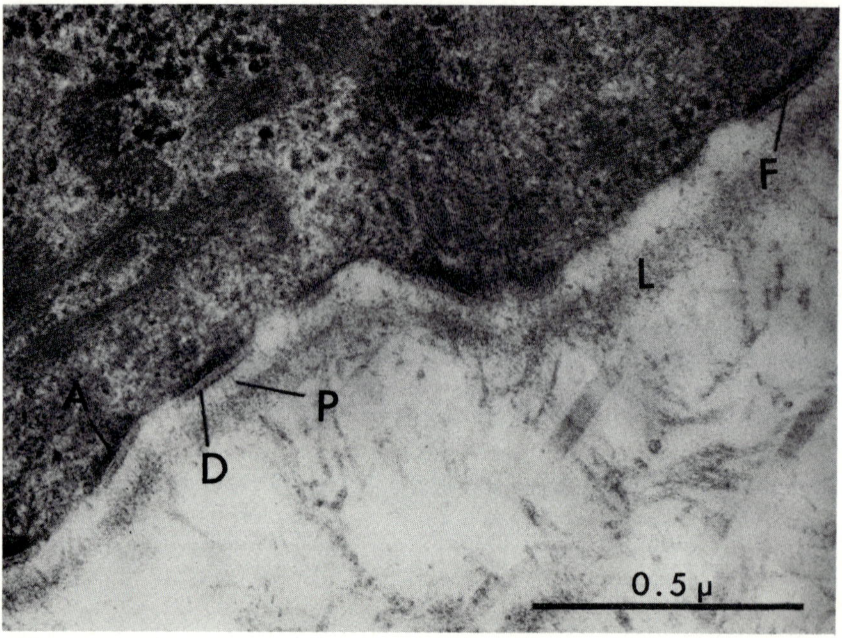

Fig. 9. Margin of a basal cell in rat gingival epithelium showing L = basal lamina; A = attachment plaque; D = peripheral density; F = proximal filaments; and P = peripheral filaments. ×72,500.

the film appears to bear a fuzzy nap often more prominent in the vicinity of cells. By a series of tests it was shown that this film consisted of substance adsorbed from serum in the medium and from materials produced by the cells. It was absent from the substratum surface when trypsin-treated and washed cells were allowed to spread in a serum free medium (Fig. 7). A characteristic of the film is that it remained tightly bound to the resin, never varying from its approximately 20 Å separation from this surface. If this is in reality a denatured protein film, its firm anchoring to the substratum is indicated by the fact that it remained in place after the detachment of cells.

The adhesion of cells, therefore, is not directly to the epoxy resin or to the tooth substance since their surfaces are actually masked by a material with physical and chemical properties different from those of the materials they cover. Possibly for this reason the presence or absence of a cuticle over the enamel appeared to have little influence on the strength of the adhesive bond between cells and the tooth.

There is evidence that cells produce, at least in part, their own surface coat. How epithelial cells synthesize and liberate this material is not known. Many of the cells examined contained rough endoplasmic reticulum whose spaces were densely packed with a granular material (see Fig. 11). There was no evidence of its being packaged in the Golgi apparatus but numerous microvesicles were seen in the cortical cytoplasm, some opening through the cell membrane onto the adhesive surface. These have been observed on basal cells by others (Odland, 1958; Hay, 1964; Schroeder and Theilade, 1966), and their function as a mechanism for exchange of materials between cytoplasm and the exterior has been suggested. The fact that the microvesicles in these cells were present on the surface applied to the impervious solid substratum rather than on surfaces exposed to nutrient fluid suggests that their function was to discharge rather than ingest materials.

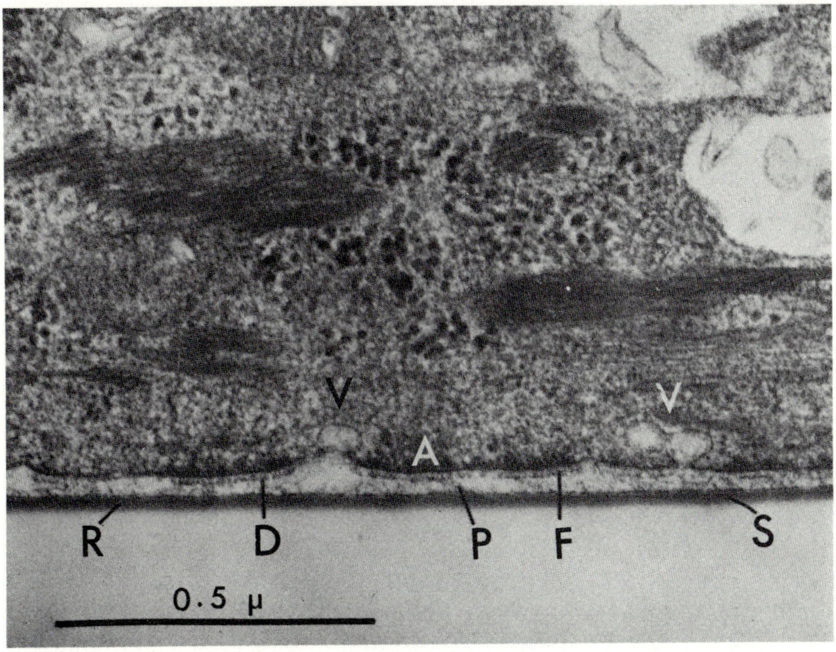

Fig. 10. Portion of a rat gingival epithelial cell cultured on epoxy resin. R = osmium stained portion of resin substratum; S = substratum film; A = attachment plaques of hemidesmosomes; F = proximal filaments; D = peripheral density; P = peripheral filaments; and V = microvesicles. ×84,700.

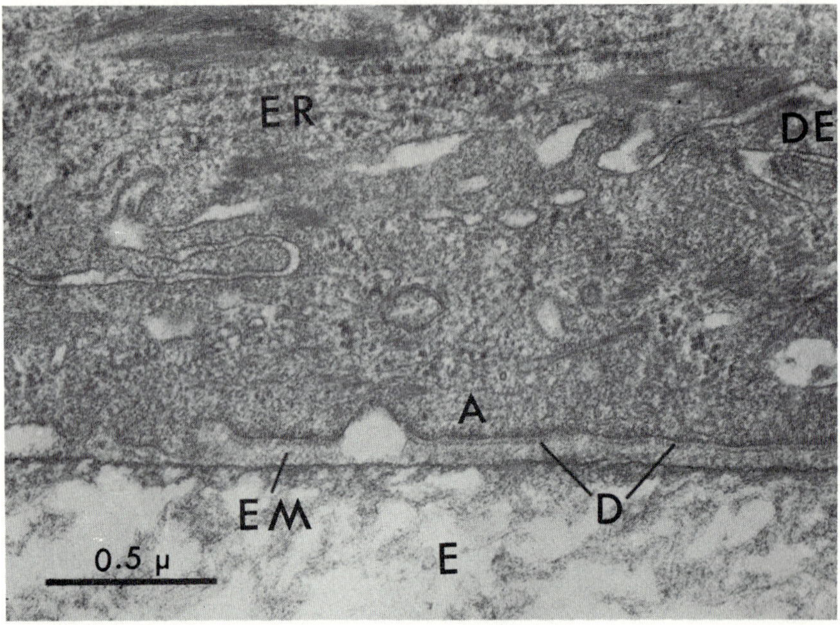

Fig. 11. Portion of a rat gingival epithelial cell cultured on rat tooth enamel. E = enamel matrix; A = attachment plaque of hemidesmosome; D = peripheral density; EM = extracellular material; DE = desmosome; and ER = endoplasmic reticulum. ×47,700.

It has been indicated in several recent studies that epithelial cells may participate in the elaboration of collagenlike molecules. Kurtz and Feldman (1962) have presented evidence that mouse kidney epithelial cells engaged in the production of the basement membrane of the glomerulus. Using labeled precursors, Hay (1964) traced by electron microscopic autoradiography the synthesis and secretion of a proline-rich protein by epidermal basal cells in a salamander. The evidence suggests that these epithelial cells, at least, in part produce the basal lamina to which they are attached.

The presence of collagen between the gingival cuff cells and the tooth crown has been verified by Stallard, Diab, and Zander (1965). Tritiated poline which they injected into mice was at first found to be incorporated within the cuff epithelium cells and later traced to the surface of the enamel. The complex nature of this material is indicated in the finding by Toto and Sicher (1964) that mucopolysaccharides are also in the same region.

Our observations on cells in culture, where contribution of materials from other cells was excluded, provide further evidence that the epithelial cell is capable of producing materials, which may become organized, as we shall see, into effective adhesive mechanism for the attachment of cells to solid surfaces.

C. An Attachment Mechanism

A characteristic feature of epithelial tissue in the organism is the specialization of certain portions of cells of the surface membrane for adhesion or attachment. The morphology of desmosomes, which connect cell with cell, and of hemidesmosomes, which hold epithelial cells to the noncellular basal lamina, has been described in detail by Odland (1958) for human epidermis and by Stern (1965) for rat oral mucosa. Presence of hemidesmosome-like structures between gingival epithelial cells and the cuticle of tooth enamel in man has been reported by Listgarten (1966), and more recently by Ito *et al.*, (1967). Unlike hemidesmosomes elsewhere in the body these are applied to a nonvital solid and in this respect would correspond to attachment mechanisms which might be found between cultured cells and their substratum.

When cells are transferred into tissue culture environment, desmosomes are not seen on the connective tissue cells. They may persist in epithelial cells where these retain contact with a neighboring epithelial cell but generally disappear quickly on exposed surfaces. No attachment bodies between either type of cell and the substratum surface have been reported.

In the electron micrographs of connective tissue and strain cells in our cultures the extraneous coat material was relatively inconspicuous. It was detectable only at points of contact between cells, or between the cell and substratum, where it formed a feltlike mass connecting the two surfaces (Figs. 6, 7). On the other hand, in the space between primary epithelial cells and substratum, the extracellular material was much more abundant and appeared to adhere upon both surfaces as an irregular tufted pile or nap (E in Fig. 8). Where the surfaces approached closely enough to each other for the free ends of their pile fibrils to meet, they appeared to join into linking strands connecting the cell membrane and the substratum film (C in Fig. 8). Over parts of the adhering cell surface, however, the material assumed a special organization resembling the extracellular portion of hemidesmosomes. The resemblance is clearly brought out by a comparison of two micrographs. In Fig. 9 is shown the portion of a cell from the normal gingiva of a rat attached to the basal membrane of the lamina propria by several hemidesmosomes whose structural features are easily recognizable. Figure 10 shows the attachment to epoxy resin of a cultured gingival epithelium cell fixed after 5 days of incubation. In both of these examples it is possible to identify the following features of a hemidesmosome: (a) the attachment plaque which is a condensation within the cytoplasm lying against the cell membrane; (b) an array of proximal filaments extending outward from each plaque area for about 140 Å where they terminate; (c) the peripheral density, a dense band appearing to be formed by the coalescence of globular masses; and (d) the peripheral filaments which extend from the peripheral density to the basal lamina or substratum film. The major difference between the hemidesmosomes in the two cases apparently lies in the degree of differentiation of the component parts of the structures, plus, of course, the fact that the basal lamina in the intact gingival mucosa is replaced by the substratum film under the cultured cells. One may, therefore, conclude that the cultured primary epithelium cells in these experiments were held to the plastic substratum by adhesive mechanisms similar in essential features to the hemidesmosomes that anchor gingival epithelial cells to underlying connective tissue.

It will be recalled that, with our jet stream detachment methods, no distinction could be made between the strength of adhesion of epithelial cells to an epoxy resin substratum and to the enamel surface of teeth in culture. When the electron micrographs of the contact surfaces in these two cases were compared (Figs. 10 and 11), the presence of hemidesmosome-like structures was apparent in both. A noticeable difference between them, however, was the greater amount of extracellular material present in the space between the cell and enamel (Fig. 11) giving to this area an electron density almost equal to that of the cell cytoplasm and making it difficult to visualize clearly the organization of the constituent materials. There does not appear to be a true basal lamina in this space.

Although it is generally assumed that hemidesmosomes act as mechanisms for adhesion of cells to surfaces, nothing is known about the strength of the attachment they afford. The fact that similar structures have been found between the crevicular epithelium and the tooth surface tells very little about how firmly the cuff cells are anchored to the tooth. If we may assume that the strength of gingival cell adhesion is comparable to that exhibited by the epithelial cells cultured *in vitro*, their attachment would be far too weak to be detected by manual probing, or even by insufflation. Such an attachment might, nevertheless, be functionally significant if it could prevent the invasion of fluid-borne foreign particles into crevicular and subepithelial spaces. The epithelial cell adhesion as observed in these experiments seems to be well adapted for such a function.

The inability of this adhesion to withstand more violent stresses, such as those to which gingivae might be subjected in the mouth, appears to be compensated for by a feature of the cell behavior observed *in vitro*. Time lapse motion picture records of cells attached to glass and plastic showed that they retained their capacity for free movement over the substrate surface. This could only mean that the bonds linking molecules of the adhesive material to each other or to the surfaces of the adherends are sufficiently labile to be broken and remade, presumably under the control of local changes in the cell surface, such as liberation of enzymes or even alteration of pH. It would imply that reattachment of epithelial cells to the tooth could be readily achieved, at least under favorable conditions. Furthermore, a dynamic attachment mechanism such as that observed in the cultured cells would be required by the gingival cuff cells in the mouth in order to provide for the movement over the enamel surface associated with their normal cycle of replacement.

VI. Questions Still Unanswered

Many obvious questions about these cell-to-surface relationships remain unanswered. A much clearer understanding will surely come when more definite information is obtained on the chemical nature of the interposed adhesive materials and particularly on the molecular configuration of their structured components.

Certain other questions also suggest themselves. Regarding the proximal filaments, we should like to know more exactly what their relation is to the unit membrane of the cell surface. Could these be extensions of the cell membrane itself, or may they be filaments extruded through openings in the membrane induced possibly by the presence of tonofibrils or the attachment plaque? What is the significance of the peripheral density? Could this be the area of junction between filaments attached to the cell membrane and other filaments anchored to the substratum, as suggested by Kelly (1966)? If so, when a cell migrates over the substratum surface, is this junction dissolved, or is the plane of separation located elsewhere? We should like information on the role of the substratum film. How universal is its occurrence? Is the firm adhesion to the substratum which it exhibits due primarily to its intimate spatial relationship with this substance, or are chemical bonds involved?

VII. Conclusions

The observations discussed have emphasized again the importance of extracellular materials in cell adhesion, and particularly in long-term adhesion to solid surfaces. It adds evidence that epithelial cells may be involved in the synthesis of this material and in its organization into mechanisms of attachment similar to hemidesmosomes. They have, more importantly, indicated the feasibility of studying these attachment mechanisms as they are formed by cells in tissue culture where the behavior of the living cells can be visualized and their adhesive reactions directly evaluated. There is a distinct advantage in tissue culture techniques: They offer a more completely defined environment for the cells and the possibility of selectively varying a number of parameters.

Acknowledgments

Supported in part by USPHS Grants 5 R01 DE 02400 and DE 02232 of the National Institutes of Health, Bethesda, Maryland.

References

Ambrose, E. J. (1961). *Exptl. Cell Res.* Suppl. 8, 54.
Baume, L. J. (1953). *J. Periodontol.* **24**, 99.
Bennett, H. S. (1963). *J. Histochem. Cytochem.* **11**, 14.
Coman, D. R. (1961). *Cancer Res.* **21**, 1436.
Curtis, A. S. G. (1962). *Biol. Rev. Cambridge Phil. Soc.* **37**, 82.
Dan, K. (1947). *Biol. Bull.* **93**, 274.
Easty, G. C., Easty, D. M., and Ambrose, S. J. (1960). *Exptl. Cell Res.* **19**, 539.
Gottlieb, B. (1921). *Deut. Monatsschr. Zehnheilk.* **39**, 142.
Greulich, R. C. (1962). *IADR Abstr.* **40**, 80.
Hay, E. D. (1964). *In* "The Epidermis" (W. Montagna and W. C. Lobitz, Jr., eds.), p. 97. Academic Press, New York.
Ito, H., Enomoto, S., and Kabayashi, K. (1967). *Bull. Tokyo Med. Dental Univ.* **14**, 267.
Kelly, D. E. (1966). *J. Cell Biol.* **28**, 51.
Kurtz, S. M., and Feldman, J. D. (1962). *J. Ultrastruct. Res.* **6**, 19.
Listgarten, M. A. (1966). *A. J. Anat.* **119**, 147.
Moscona, A. (1961). *Nature* **190**, 408.
Odland, G. F. (1958). *J. Biophys. Biochem. Cytol.* **4**, 529.
Orban, B. (1949). "Oral Histology and Embryology." Mosby, St. Louis, Missouri.

Parfitt, G. J. (1961). *J. Dental Res.* **40**, 705.
Pethica, B. A. (1961). *Exptl. Cell Res.* Suppl. 8, 123.
Rambourg, A., Neutra, M., and Leblond, C. P. (1966). *Anat. Record* **154**, 41.
Rappaport, C., and Bishop, C. B. (1960). *Exptl. Cell Res.* **20**, 580.
Rinaldini, L. M. (1959). *Expt. Cell Res.* **16**, 477.
Rosenberg, M. D. (1960). *Biophys. J.* **1**, 137.
Rubin. H. (1966). *In* "Major Problems in Developmental Biology" (M. Locke, ed.), p. 315. Academic Press, New York.
Schroeder, H. E., and Theilade, J. (1966). *J. Periodontol. Res.* **1**, 95.
Sicher, H. (1962). "Orban's Oral Histology and Embryology." Mosby, St. Louis, Missouri.
Skougaard, M. R., and Beagrie, G. S. (1962). *Acta Odontol. Scand.* **20**, 467.
Stallard, R. E., Diab, M. A., and Zander, H. A. (1965). *J. Periodontol.* **36**, 130.
Stern, I. B. (1965). *Periodontics* **3**, 224.
Taylor, A. C. (1961). *Exptl. Cell Res.* Suppl. 8, 155.
Toto, P. D., and Sicher, H. (1964). *Periodontics* **2**, 154.
Waerhaug, J. (1952). *Odontol. Tidskr.* **60**, Suppl. 1.
Weiss, L. (1960). *Intern. Rev. Cytol.* **9**, 187.
Weiss, L. (1961). The measurement of cell adhesion. *Exptl. Cell Res.* Suppl. 8, 141.
Weiss, L. (1962). *J. Theoret. Biol.* **2**, 236.
Weiss, P. (1958). *Intern. Rev. Cytol.* **7**, 391.
Zander, H. A. (1956). *J. Dental Res.* **35**, 308.

CHAPTER 4

ATTACHMENT OF MARINE BACTERIA TO SOLID SURFACES

William A. Corpe

The development of bacterial populations in natural environments, such as lakes, ponds, soil, seawater, and sediments, occurs most abundantly on the surfaces of solid particles or films. Dead animal or plant cells and their derivatives, as well as surfaces of inorganic particles, can become populated in a short time by bacteria that cannot be removed by simple washing. The growth of microorganisms on submerged surfaces is known as microbial fouling. Although it occurs in fresh water, its main importance is in brackish water and marine environments, where microbial slimes may become precursors of heavy fouling growths of macroorganisms (Wood, 1967).

I. Observation of Attached Bacteria

The Cholodny (1930) technique, originally employed for direct study of soil bacteria, has become a traditional method for study of aquatic bacteria as well, since glass slides can be examined by the microscope and cultures can be isolated from the surfaces. Henrici and Johnson (1935) adopted the method for study of attached bacteria in fresh water lakes, and studies by ZoBell (1946) and Kriss (1963) showed that bacteria and other microorganisms became attached to glass slides at many depths in several ocean regions. Zvyagintsev (1959) showed great variation in capacity of bacteria to adsorb to glass; some adsorbed very strongly, others not at all.

Meadows and Anderson (1966, 1968) have carried out with considerable success the direct microscopic examination of microorganisms attached to marine sand grains. They observed stained microcolonies of bacteria, blue-green algae, and diatoms that suggested growth of organisms settled on the surface. Early studies by ZoBell (1943) demonstrated quite clearly that bacteria which attach to solid surfaces actually required them for growth, especially when the nutrient level in the aquatic environment was quite low. He emphasized that attachment of cells is not a passive phenomenon but rather one that requires a period of incubation with nutrients, indicating that some synthesis may be necessary. Little or no attempt has been made to investigate the mechanism of adhesion by bacteria to solid substrata although references to sticky surfaces, holdfasts, surface charge, and so on are common in the literature. Bacterial cells grown in natural environments on slides often look quite different from cells grown on laboratory media. Morphological differences may reflect cytological and compositional differences as well. This possibility must be taken into account when studying attachment of bacteria to solid surfaces.

Whether the bacteria and other microscopic forms concerned with development of primary films merely settle fortuitously (Floodgate, 1966) or are attracted to a nutrient-enriched surface by chemotaxis, a phenomenon shown by Adler (1966) to occur in bacteria has not been determined. The opportunity for growth or metabolism seems to be critical in determining whether a bacterium will become firmly attached to a surface or not (Heukelikian and Heller, 1940; ZoBell, 1943).

One can rinse slides, with attached bacteria, at least five times with flowing, sterile seawater, without any substantial change in the attached microflora; however, the total microscopic count per cm^2 of surface is consistently higher than the bacterial plate count per cm^2. If slides were rinsed with tap water or distilled water, a substantial loss in microscopic count would occur.

The series of photographs in Fig. 1 shows the attached bacterial population after 24 (a–e), 48 (f–h), and 72 hours (l–p) in seawater. These photographs are like others taken from slides exposed to natural environments (Henrici and Johnson, 1935; Kriss, 1963) in that they show cells on filaments and stalks, in masses or short chains, seemingly undergoing cell division. Close examination reveals some change in population with an increase in the exposure time (Fig. 1).

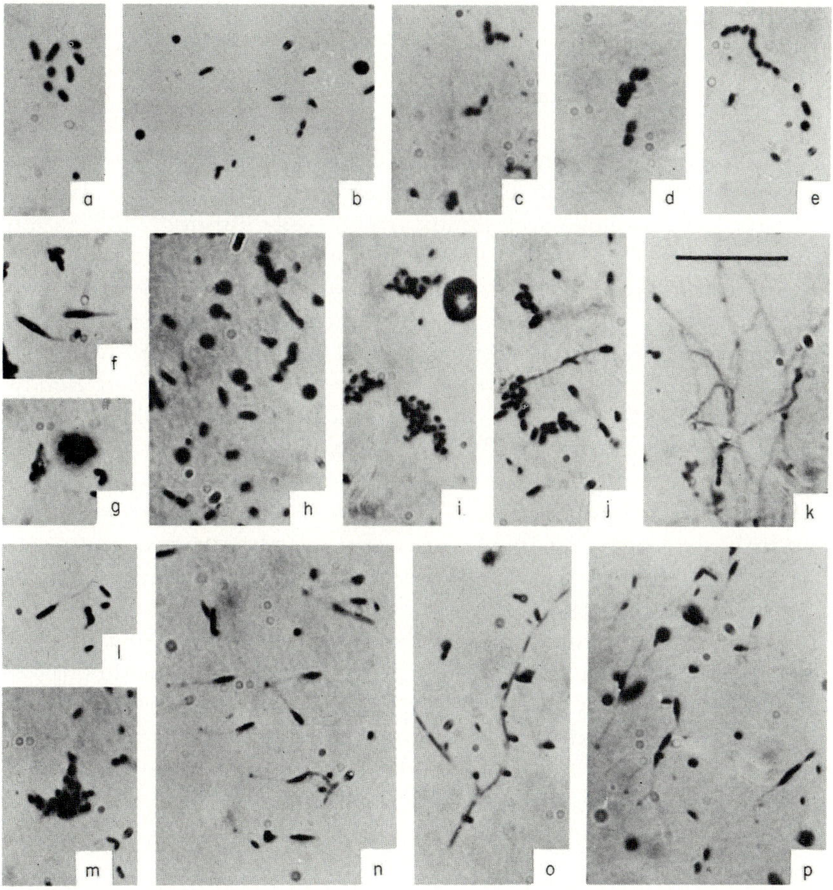

Fig. 1. The attachment of bacteria to glass slides. Slides were submerged in a regulated marine aquarium, fixed with 2% acetic acid and stained with Huckers crystal violet solution. (a–e) from slides held for 24 hours, (f–h) from slides submerged for 48 hours, and (l–p) from slides submerged for 72 hours. The scale shown is 5 μ.

The scanning electron microscope shows great promise in providing a means for study of the surface character of solid substrata (Kimoto and Russ, 1969); it is certain to allow a detailed study of individual cells and microcolonies of attached bacteria for adhesive organelles and details of cell surface structure. Cytologically demonstratable surface components such as fimbriae, capsules, slime, and cements which may occur on the bacterial surface largely external to the cell wall must obviously be evaluated for adhesive properties. Salton (1964) recognizes the "ionic layer" of the bacterial cell which affects the overall surface charge that in turn may influence attachment of the bacteria to solid surfaces.

II. Surface Charge of Bacterial Cells

Bacterial cells have a net negative charge to which surface components contribute. Although electrophoretic mobilities of bacteria have been studied (Gebicki and James, 1962), only limited information about the precise nature of surface components has been provided. Douglas and Parker (1958) have shown there was a change in electrophoretic mobility when *Bacillus megatherium* cells were treated with lysozyme which suggests that alteration of surface structure can affect the electrical charge of the cells. More recently Neihof and Echols (1968) examined the electrophoretic mobilities of cell walls of a number of microorganisms. Walls were dominated by electronegative charges, and mobilities were very similar to those of the corresponding intact cells. The nature of the negatively charged groups has not been fully explored, but they seem to be mainly carboxyl and phosphate groups. Removal of 98% of the phosphate from walls of *Bacillus megatherium* by treatment with phosphatase and extraction with trichloroacetic acid changed the mobility-pH behavior of the walls, but they still carried strong acid groups, presumably due to residual phosphate (Echols and Neihof, 1969). Charge interaction between cells and surfaces may lead to tight binding of cells to material like clay (Santoro and Stotzky, 1968) and sand (Meadows and Anderson, 1968). The experimental use of ion exchange resins to examine differences in affinity by microorganisms has been suggested by the work of Zvyagintsev (1962).

Bacterial cells may be induced to synthesize new surface components, and by changing the environmental circumstances, mutants may be selected which display a different overall surface charge. "Adaptive" structural properties recognized in bacteria that may be related to the organism's propensity for attachment include holdfasts, surface slime, and adhesive pili. The composition and physical nature of these structured and unstructured materials as well as the more strictly constitutive parts of the cell surface must be considered in the general discussion of bacterial adhesion to solid surfaces.

III. Fimbriae or Pili

Long, straight, thin (100 Å), hairlike structures arranged perpendicular to the cell surface have been recognized in many species of the family Enterobacteriaceae as well as species of *Pseudomonas* and *Caulobacter* but not yet in bacteria of other groups (Duguid, 1968). The structures have been referred to as fimbriae (Duguid, 1959) or pili (Brinton, 1967). Fimbriae are a class of bacterial surface filaments, each type with a characteristic structure whose common feature is to pass through the cell envelope without altering its integrity. Type 1 fimbriae are pure proteins. They are adhesive and enable fimbriated bacteria to adhere, perhaps irreversibly, to one another, particles, glass, and cellulose surfaces (Duguid, 1959). The net surface electric-charge density of type 1 fimbriae is lower than that of the outermost cell wall layer. They contribute a large amount of surface area to fimbriated cells enabling them to adhere to solid surfaces such as dead or living animal and plant cells, when nutrients become available by lysis, leakage, or excretion. Fimbriated cells display a strong tendency to grow on the surface of static liquid media as a pellicle of aggregated cells. Fimbriae of types 2-6 (Duguid, 1968) are distinguished in part by their adhesiveness or lack of adhesiveness for a particular surface.

Type F fimbriae are the so-called sex fimbriae. They act as specific receptors for F specific phage and are thought to act as transfer tubes for injection of bacterial DNA. They are distinguishable in the electron microscope since they are thicker and longer than other types.

IV. Capsular and Slime Polysaccharides

The synthesis of capsular polysaccharides or similar sticky surface material may serve to cement the cells and their microcolonies to solid substrata. Many bacterial species release exopolysaccharides into laboratory media (Wilkinson, 1958) and, presumably, may do so in natural environments as well. Capsules and slime of bacteria may be a compositionally simple homopolymer or very complex heteropolymers containing several different sugars linked in a variety of ways, each with unique physical properties (Stacey and Barker, 1960). Cer-

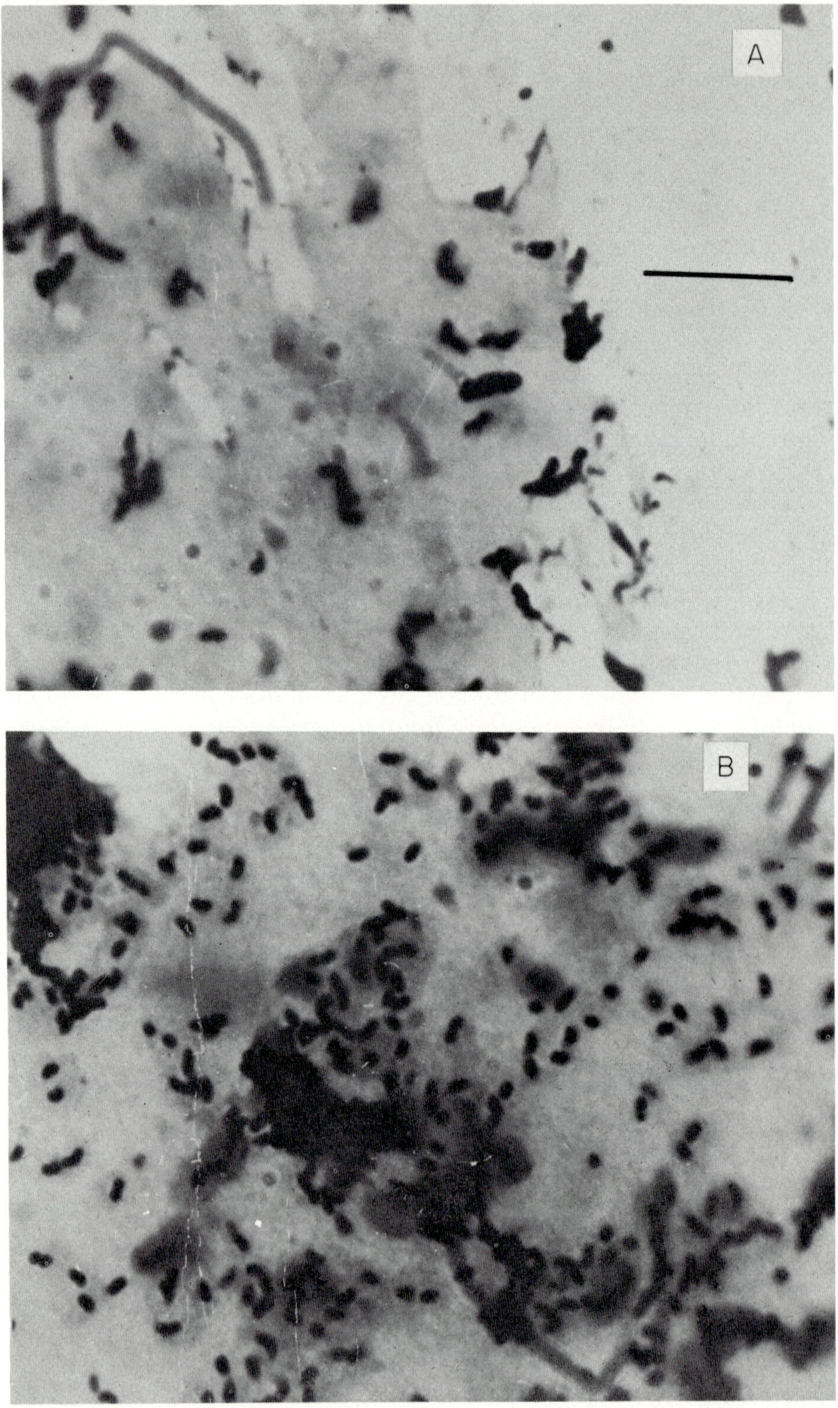
A
B

tainly if capsular polysaccharides have a different chemical character from that of the cell wall of the organisms, the overall surface charge of the cell will be modified when they are produced.

ZoBell (1943) stated that when sessile bacteria come into contact with a solid surface they secrete a cementing substance. When the cells of the organisms were mechanically removed, they left a faintly staining film. Studies of stained submerged slides by the author seemed to confirm this view. The photographs in Fig. 1 (g,m) show groups of cells, microcolonies, surrounding a lightly staining matrix. Some isolates from the marine environment have the ability to cover the cells and solid surfaces with a sheet of lightly stained polymer (Fig. 2). A polymer has been recovered from seawater-peptone broth cultures of the organism which give qualitative tests for carbohydrates.

Chapman, Murray, and Salton (1963) have described an intercellular material responsible for cementing *Lampropedia hyalina* cells into regular aggregates. The material is a high molecular weight polysaccharide similar to extracellular slime and capsules in that it is dispensible because mutant cells which have lost the ability to produce such material seem to survive without difficulty. Cells of *Sarcina ventriculi* are cemented together into regular packets with an intercalating material which has been identified as cellulose (Canale-Parola *et al.*, 1961). Microfibrils of cellulose are laid down by *Acetobacter xylinum* (Mühlethaler, 1949) in such a manner that the cells become enmeshed, yielding a tough pellicle. Deposition of extracellular polysaccharides as a highly cross-linked mesh has been also observed in *Chromobacterium violaceum* (Corpe, 1964). The polysaccharide in the latter case, however, is a complex material containing glucose and amino sugars. Cohesive polymers with a similar function but different composition have been isolated from several marine bacteria (Corpe, 1969). The colonies of these organisms like those of *Chromobacterium violaceum* adhere to agar and other "open" surfaces, but not to glass or plastic.

Fig. 2. Pure cultures of *Pseudomonas atlantica* (Humm), strain 4b, growing as an attached film on the surface of a glass slide. The slide was suspended in a culture of artificial seawater containing 0.01% (w/v) Bacto-peptone. The film was stained with Huckers crystal violet following fixation with 2% acetic acid. In A, the film was scraped or torn to better show the contrast between the lightly stained film and the unstained glass beneath. Photograph B shows that the filming polymer has greater density around cell masses. The scale shown is 5 μ.

It has been suggested (Martin and Richards, 1963) that in comparison with other organic substances, polysaccharides bind soil particles most effectively into aggregates. Some polymers produced by microorganisms are quite resistant to decomposition (Corpe, 1960; Martin and Richards, 1963), which makes reasonable the observation that polysaccharides constitute a stable proportion of soil humus, and suggests that a portion of the polymers may be of microbial origin (Whistler and Kirby, 1956). Similar studies have not been made of marine sediments; moreover, almost no information is available regarding the chemical and physical nature of polysaccharides from marine bacteria.

V. The Bacterial Cell Wall (Envelope)

The rigid framework of the cell wall of bacteria has as its main function the protection and support of the labile portions of the cell, the plasma membrane, and the cytoplasm. The envelopes of gram-negative bacteria, the predominant type found in the sea (ZoBell, 1946), are composed mainly of protein-lipid and polysaccharide which, because of its chemical and physical character, may be sticky or tacky. The cell envelopes have two lipoprotein membranes separated by a thin layer of glycosamino-peptide which is responsible for cell shape and rigidity (Murray *et al.*, 1965). Isolated envelopes of gram-negative are composed of protein (35-84%), lipid (11-25%), and polysaccharide (5-10%), and about 3-10% glycosaminopeptide (Salton, 1964). Polysaccharides from cell walls of many gram-negative bacterial species have been studied (Davies, 1960) and have been found to contain mixtures of sugars representing most known classes such as pentoses, methyl pentoses, unusual heptoses, as well as several amino sugars not previously known (Salton, 1965).

Cell envelopes from a few marine bacteria (Smithies *et al.*, 1955; Brown and Shorey, 1963; Sud and Tyler, 1964; Buckmire and MacLeod, 1965) have been isolated and studied. There has been no effort to study the composition or surface structure of envelopes of bacteria with a propensity for attachment and to compare them with those of bacteria that do not attach.

The topography of some gram-negative bacterial cells is remarkably convoluted (Murray, 1963; Murray *et al.*, 1965). The convolutions seem to be restricted to the outer superficial lipid-protein-polysaccharide membrane. If the outer layer of the cell were plastic in the living state, one might be tempted to suspect it could spread out and flow into or otherwise accommodate to solid surfaces. This, of course, would increase the opportunity for surface forces to act and, thus, to facilitate adhesion.

The physical and chemical properties of bacterial cell surfaces are not only of importance for attachment to solid substrata but also must be related to tendency of cells to aggregate into clumps or regular arrangements (Hoffman, 1964).

VI. Bacteria with Stalks and Holdfasts

Henrici and Johnson (1935) described a number of bacteria which attached to the surfaces of glass slides suspended in freshwater lakes; some of them were stalked. In the 30 years that have passed since the studies of Henrici and Johnson were completed, several species of stalked caulobacters have been isolated and studied. Recently, an extensive study of the group by Poindexter (1964) has appeared. Electron micrographs of thin sections of the stalked cells showed that the wall of the stalk is continuous with the cell wall and is not a secreted material as once imagined. At the end of the stalk is a holdfast material which apparently *is* secreted and by which the attachment of the organism is accomplished.

The adhesive holdfast permits attachment almost irreversibly to a variety of solid substrata (glass, collodion, cotton fibers) to one another and to other cells. The chemical nature of the holdfast material is still unknown, but according to Poindexter (1964), is neither carbohydrate nor protein. Methods used to resolve this point, however, were not exhaustive.

Hyphomicrobium vulgare produces hyphal filaments, the distal end of which swells and forms an oval bud which grows until it reaches the size of the mother cell. The cells sometimes become arranged into rosettes with the hyphae turned out (Hirsch and Conti, 1964). *H. vulgare* has been isolated from a variety of natural sources including seawater (Guillard and Watson, 1962). Leifson (1964) isolated a species from a sample of seawater which he called *H. neptunicum.* Conti and Hirsch (1965) have shown an electron micrograph of a holdfast of a *Hyphomicrobium* species. Nothing is known about their chemical nature or physical properties.

There is an apparent morphological relationship between *Hyphomicrobium* and *Caulobacter* species (Starr and Skerman, 1965) in view of the fact that both hypha and stalk are living parts of the cell and not secretions as once believed. Both of these genera form a major part of the attached bacterial flora in seawater (Figs. 1 and 3). Other stalked

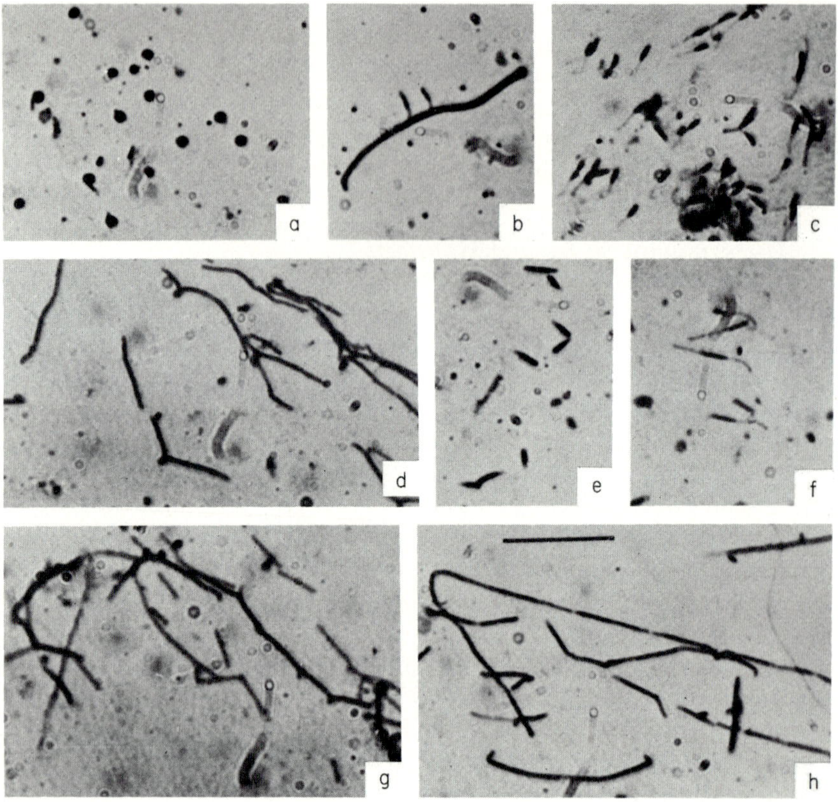

Fig. 3. Stalked and filamentous bacteria with holdfast structures attached to glass slides. The photographs were taken from stained slides which had been submerged in the sea at San Diego for 96 hours. The scale shown is 5 μ.

bacteria which seem to attach by holdfasts are the genera *Nevskia* and *Blastocaulis* (Skerman, 1967).

The participation of *Hyphomicrobium* species in the deposition of manganese in freshwater pipeline has been studied by Tyler and Marshall (1967a, b). The phenomenon of deposition of iron and manganese by budding, sheathed, and stalked bacteria will be discussed below in connection with attachment of microorganisms by inorganic cements.

Filamentous sheathed bacteria of the genus *Sphaerotilus* and other Chlamydobacteriales (Mulder, 1964) show the ability to attach to solid substrata (Pringsheim, 1949). The presence of specific holdfasts, the sticky character of the sheath, and perhaps the ability of the organisms to "grow into" imperfections of a rough surface may all contribute to their propensity for attachment. At least eight genera of the order *Chlamydobacteriales* are described as possessing holdfasts (Skerman, 1967).

Gliding bacteria of the genus *Saprospira* described by Lewin (1965) in natural environments may become attached to glass slides and show long filaments with "knees" and what appear to be holdfast structures (Lewin, 1968). Photomicrographs in Fig. 3 show these organisms attached to glass slides submerged 96 hours in the sea from Scripps pier in La Jolla.

There are examples in other bacterial genera of organisms which secrete holdfast material from a specific area on the cell surface. The genus *Blastobacter* described by Zavarzin (1961) has pear-shaped cells attached to the substrate directly by a holdfast located at the narrow end. The adhesive holdfast of the genus *Asticcacaulis* forms at the pole of the cell and next to the excentral stalk (Poindexter, 1964). Brock (1968) suggests that a large portion of the bacteria epiphytic on aquatic plants are composed of groups which possess holdfast structures.

VII. Attachment by Deposition of Inorganic "Cements"

A number of bacterial groups are able to oxidize ferrous iron to the ferric form; they include sheath forming species of the *Spherotilus-Leptothrix* group, stalked bacteria of the genus *Gallionella* and certain species of the genus *Thiobacillus*. Iron bacteria of the *Spherotilus-Leptothrix* group also participate in the oxidation of manganous compounds to give MnO_2 precipitates (Mulder, 1964; Pringsheim, 1949; Silverman and Ehrlich, 1964; Wolfe, 1964). Cell surfaces presumably have an affinity for iron and manganese which become fixed as hydroxides or salts. Ferric hydroxide or manganic oxide precipitates may be laid down within the matrix of the *Leptothrix* sheath or deposited on the surface of the sheath or cells. Recently Tyler and Marshall (1967a, b) presented evidence that stalked, budding bacteria of the genus *Hyphomicrobium* inhabiting freshwater pipelines deposit manganese oxide in the network of branching filaments. The binding or encircling of cells or their residues into a union with a solid substratum causes one to view the deposits as mineral cements.

Greenfield (1963) has described bacteria isolated from seawater with the ability to precipitate carbonates as insoluble calcium or magnesium salts. The precipitation occurs in seawater peptone broth and the voluminous crystals are cemented rather firmly to the sides of culture vessels (Corpe, 1969). The deposition of insoluble carbonates is dependent on the enzymic production of CO_2 and NH_3 from amino acids (Silverman and Ehrlich, 1964) and is accompanied by an increase in alkalinity of the medium. According to Greenfield (1963) the cells which cause the precipitation become trapped within the crystals. Certainly the crystalline mass is thoroughly infused with cells and growth products, and it seems entirely possible that the cells or products may not only be passively trapped by crystals but in fact may contribute to the tendency of the crystalline mass to adhere to the surface of the culture vessel.

VIII. The Adhesiveness of Bacteria in Relation to the Nature of the Substratum

A major requirement for adhesion of a bacterium to a substratum is an intimate contact of some portion of the bacterial surface with the substratum. The amount of contact and the strength of adhesion may depend in part upon the structure and composition of the bacterial surface. We know something about the structure and ultrastructure of the surfaces of a few species of bacteria, but the relationship between the attachment propensity of cells and the character of their surfaces is obscure. Surface irregularities of the substratum, of course, might increase the opportunity for contact, but the physical and chemical character of substratum is perhaps the most important in determining which organisms will be involved and how firm the attachment becomes. Other environmental conditions such as pH, oxidation-reduction potential, temperature, presence of growth inhibitory, and lytic agents all affect the microenvironment where attachment occurs (Alexander, 1964). These conditions would be expected to affect the species which attach and perhaps the order of their appearance as well. ZoBell (1970) has described in some detail the substratum as it influences the growth and development of aquatic microorganisms.

IX. Relationship of Microbial Films to Marine Fouling

The adhesion of bacteria and other microorganisms and their products to solid surfaces may affect the settlement of larvae of marine animals (Crisp and Ryland, 1960; Meadows and Williams, 1963) and thereby serve as precursor of heavy destructive fouling or corrosion (ZoBell, 1939). A discussion of microbial film formation in relation to marine fouling is presented by Wood (1967) who indicates there is some disagreement among researchers concerning the relative importance of bacteria and microalgae in the development of slimy films on surfaces exposed to seawater. In some environments diatoms have been reported to be present in great abundance and bacteria scarce. In other environments the opposite seems to be true. In our own observations, bacteria were the first organisms to appear on the slides exposed to waters facing the open sea. Diatoms did not appear in more than an occasional microscopic field until after 96 hours. At no time during a two-week period of submergence did the numbers of diatoms or other algae exceed that of bacterial cells or, for that matter, microcolonies of bacteria (Corpe, 1970).

In polluted waters the situation is somewhat confused, mainly because events leading to primary film formation are greatly accelerated. The abundance of a great variety of soluble and particulate organic substrata gives rise to a very complex population of bacteria, protozoa, and diatoms (Corpe 1969), making the sequence of events leading to development of primary film very difficult to sort out.

Acknowledgments

Studies on attachment of bacteria to surfaces in the author's laboratory are supported by contract N00014-68-A-0025-0001 from the Office of Naval Research.

I wish to acknowledge Louise L. Slade for her assistance in the laboratory, and Dr. C. E. ZoBell for some helpful discussions.

References

Adler, J. (1966). *Science* **153**, 708.
Alexander, M. (1964). *Ann. Rev. Microbiol.* **18**, 219.
Brinton, C. C., Jr. (1967). *In* "The Specificity of Cell Surfaces" (B. D. Davis and L. Warren, eds.), pp. 37-70. Prentice-Hall, Englewood Cliffs, New Jersey.
Brock, T. D. (1968). "Principles of Micobial Ecology." Prentice-Hall, Englewood Cliffs, New Jersey.
Brown, A. D., and Shorey, C. D. (1963). *J. Cell Biol.* **18**, 681.
Buckmire, F. L. A., and MacLeod, R. A. (1965). *Can. J. Microbiol.* **11**, 677.
Canale-Parola, E., Borasky, R., and Wolfe, R. S. (1961). *J. Bacteriol.* **81**, 311.
Chapman, J. A., Murray, R. G. E., and Salton, M. R. J. (1963). *Proc. Roy. Soc.* **B158**, 498.
Cholodny, N. (1930). *Arch. Mikrobiol.* **1**, 620.
Conti, S. F., and Hirsch, P. (1965). *J. Bacteriol.* **89**, 503.
Corpe, W. A. (1960). *Can. J. Microbiol.* **6**, 153.
Corpe, W. A. (1964). *J. Bacteriol.* **88**, 1433.
Corpe, W. A. (1969). Unpublished observations.

Corpe, W. A. (1970). *In* "Developments in Industrial Microbiology" (C. J. Corum, ed.) Vol. II, Chap. 39, Society for Industrial Microbiology, Washington, D.C.
Crisp, D. J., and Ryland, J. S. (1960). *Nature* **185**, 119.
Davies, D. A. L. (1960). *Advan. Carbohydrate Chem.* **15**, 271.
Douglas, H. W., and Parker, F. (1958). *Biochem. J.* **68**, 99.
Duguid, J. P. (1959). *J. Gen. Microbiol.* **21**, 271.
Duguid, J. P. (1968). *Arch. Immunol. Therap. Exptl.* **16**, 173.
Echols, W. H., and Neihof, R. A. (1969). *Bacteriol. Proc.* p. 49.
Floodgate, G. D. (1966). *Verhandl. Inst. Meeresforsch. Bremerhaven, Sond. II.* p. 265.
Gebicki, J. M., and James, A. M. (1962). *Biochim. Biophys. Acta* **59**, 167.
Greenfield, L. H. (1963). *Ann. N.Y. Acad. Sci.* **109**, 23.
Guillard, R. R., and Watson, S. W. (1962). *Oceanas* **8**, 22.
Henrici, A. J., and Johnson, D. E. (1935). *J. Bacteriol.* **30**, 61.
Heukelikian, H., and Heller, A. (1940). *J. Bacteriol.* **40**, 547.
Hirsch, P. A., and Conti, S. F. (1964). *Arch. Mikrobiol.* **48**, 339.
Hoffman, H. (1964). *Ann. Rev. Microbiol.* **18**, 111.
Kimoto, S., and Russ, J. C. (1969). *Am. Scientist* **57**, No. 1, 112.
Kriss, A. E. (1963). "Marine Microbiology" (Deep Sea) (translated by J. M. Shewan and Z. Kabata). Oliver & Boyd, Edinburgh and London.
Leifson, E. R. (1964). *Antonie van Leeuwenhoek, J. Microbiol. Serol.* **30**, 245.
Lewin, R. A. (1965). *Can. J. Microbiol.* **11**, 77.
Lewin, R. A. (1968). Personal communication.
Martin, J. P., and Richards, S. J. (1963). *J. Bacteriol.* **85**, 1288.
Meadows, P. S., and Anderson, J. G. (1966). *Nature* **212**, 1057.
Meadows, P. S., and Anderson, J. G. (1968). *J. Marine Biol. Assoc. U.K.* **48**, 161.
Meadows, P. S., and Williams, G. B. (1963). *Nature* **198**, 610.
Mühlethaler, K. (1949). *Biochim. Biophys. Acta* **3**, 527.
Mulder, E. G. (1964). *Antonie van Leeuwenhoek, J. Microbiol. Serol.* **29**, 121.
Murray, R. G. E. (1963). *Can. J. Microbiol.* **9**, 381.
Murray, R. G. E., Steed, P. and Elson, H. E. (1965). *Can. J. Microbiol.* **11**, 547.
Neihof, R. A., and Echols, W. H. (1968). *Naval Research Laboratory Rept.* **6795**.
Poindexter, J. S. (1964). *Bacteriol. Rev.* **28**, 231.
Pringsheim, S. G. (1949). *Phil. Trans. Roy. Soc. London* **B233**, 453.
Salton, M. R. J. (1964). "The Bacterial Cell Wall." Elsevier, Amsterdam.
Salton, M. R. J. (1965). *Ann. Rev. Biochem.* **34**, 143.
Santoro, T., and Stotzky, G. (1968). *Can. J. Microbiol.* **14**, 299.
Silverman, M. P., and Ehrlich, H. L. (1964). *Advan. Appl. Microbiol.* **6**, 157.
Skerman, V. B. D. (1967). "A Guide to the Identification of the Genera of Bacteria," 2nd ed. Williams & Wilkins, Baltimore, Maryland.
Smithies, W. R., Gibbons, N. E., and Bayley, S. T. (1955). *Can. J. Microbiol.* **1**, 606.
Stacey, M., and Barker, S. A. (1960). "Polysaccharides of Micro-organisms." Oxford Univ. Press (Clarendon), London and New York.
Starr, M., and Skerman, V. B. D. (1965). *Ann. Rev. Microbiol.* **19**, 409.
Sud, I. J., and Tyler, M. E. (1964). *J. Bacteriol.* **87**, 696.
Tyler, P. A., and Marshall, K. P. (1967a). *Arch. Mikrobiol.* **56**, 344.
Tyler, P. A., and Marshall, K. P. (1967b). *Antonie van Leeuwenhoek, J. Microbiol Serol.* **33**, 171.
Whistler, R. L., and Kirby, R. W. (1956). *J. Am. Chem. Soc.* **78**, 1755.
Wilkinson, J. F. (1958). *Bacteriol. Rev.* **22**, 46.

Wolfe, R. S. (1964). *In* "Principles and Applications in Aquatic Microbiology" (H. Heukelekian and N. C. Dondoro, eds.), pp. 82-97. Wiley, New York.

Wood, E. J. F. (1967). "Microbiology of Oceans and Estuaries." Elsevier, Amsterdam.

Zavarzin, G. A. (1961). *Microbiology (USSR) (English Transl.)* **30**, 774.

ZoBell, C. E. (1939). Circ. 588, Sci. Sect. Natl. Paint, Varnish & Lacquer Assoc., San Francisco, California.

ZoBell, C. E. (1943). *J. Bacteriol.* **46**, 39.

ZoBell, C. E. (1946). "Marine Microbiology." Chronica Botanica, Waltham, Massachusetts.

ZoBell, C. E. (1970). *In* "Marine Biology" (O. Kinne, ed.), Vol. 1, J. Wiley, New York.

Zvyagintsev, D. G. (1959). *Microbiology (USSR) (English Transl.)* **28**, 104.

Zvyagintsev, D. G. (1962). *Microbiology (USSR) (English Transl.)* **31**, 275.

CHAPTER 5

A REVIEW OF SOME FACTORS AFFECTING SETTLEMENT AND ADHESION IN THE CYPRIDS OF SOME COMMON BARNACLES

H. Barnes

I. Introduction

When an animal is permanently attached in its adult form and yet has free-living planktonic larval stages, it must at some point in its life history become securely attached to a suitable substratum on which the adult can develop. In cirripedes this transition takes place at the cypris larval stage by the secretion of an adhesive cement. Although this final settlement is only the culmination of a series of related events, it will not be necessary in a publication specifically addressed to the topic of adhesion to deal with the factors affecting the initial attraction of the cyprid to a solid substratum.

The eggs of a typical balanid cirripede—incubated in the mantle cavity of the adult—produce a ripe embryo, set free as a free-swimming, stage I nauplius larva, which is said not to feed. There follow five further planktonic, feeding nauplii, the stage VI nauplius sometimes being referred to as a metanauplius. The successive nauplius stages, similar and characteristic in form, show a gradual increase in size and complexity—particularly obvious in the various appendages and spines which adorn the larval carapace. The stage VI nauplius gives rise to the cyprid (cypris larva)*—in general appearance quite different from the preceding nauplius stage. The cyprid is at first free-swimming but nonfeeding; when, however, it comes into contact with a solid surface, it may attach to it by the secretion of a cement and after attachment will metamorphose to the young adult. We shall be concerned only with this cypris stage and in particular with its reactions on contact with a solid surface.

II. General Morphology of the Cyprid

For the purposes of this discussion it is not necessary to consider the internal anatomy of the larva in detail; however, the discussion does require a brief description of the morphology of the "mature" larva (i.e., after the nauplius stage VI—cyprid ecdysis), with particular reference to those parts concerned in settlement.

There are a number of early descriptions of the cypris larva (Darwin, 1851, 1854; Claus, 1869; Hoek, 1884; Hoffendahl, 1904). A more detailed account based on histological work was given by Doochin (1951). The most precise account is that of Walley (1969) who confirmed and extended Doochin's observations; Walley's account of the cyprid of *Balanus balanoides* will be closely followed.† The whole larva is lodged within a bivalve carapace (Fig. 1), the two halves of which may be drawn together by a prominent adductor muscle. An anterior and posterior cavity are separated by the oral cone which bears the rudimentary adult mouth parts. The thorax lies within the posterior mantle cavity. The organs with which we shall be particularly concerned, the antennules, arise dorsolaterally in the anterior mantle cavity.

*The larva is properly termed the cyrpis larva; the organism is commonly referred to as a cyprid; this term will be used although "cypris" will be retained for any adjectival use.

†Dr. J. Rees has very kindly allowed me to read the manuscript of a paper in press (Walley, 1969) on the larval stages and early adult of *Balanus balanoides*. Dr. Nott and Dr. Foster (1969) have provided a proof copy of their paper on an electron microscope study of the antennule of the same species. The account given here owes much to their papers, and it is to be hoped that any similarity in wording will be taken as a compliment to their clarity of expression upon which it would be difficult to improve.

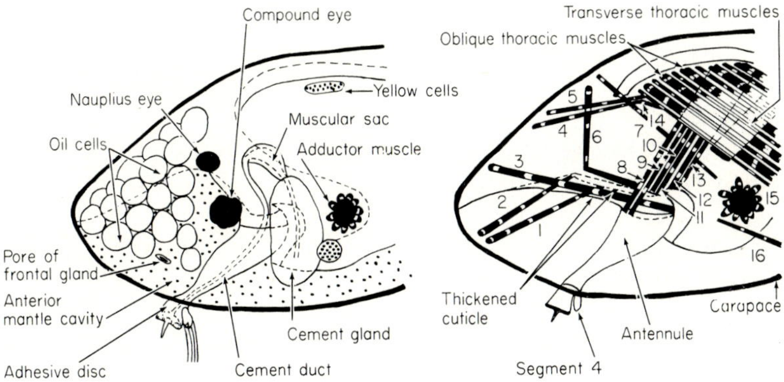

Fig. 1. Diagram of anterior part of cyprid of *Balanus balanoides* showing antennule and other parts mentioned in text. After Walley, 1969.

Proximally, the cuticle of each antennule, which is made up of four segments, is laterally thickened to form an anterior–posterior rod. The proximal, first segment is directed to the posterior in such a way that it articulates with the next and anteriorly directed segment in a distinct "elbow" accommodated in a posterior pocket of the anterior mantle cavity. This second segment is long and tapers to its junction with the third, bell-shaped adhesive segment (see p. 93) from which the fourth and most distal segment arises laterally. The cuticle on the lower "adhesive" surface of the antennal disc has a felt of short hair-like processes first emphasized by Nilsson-Cantell (1921) (see p. 93);

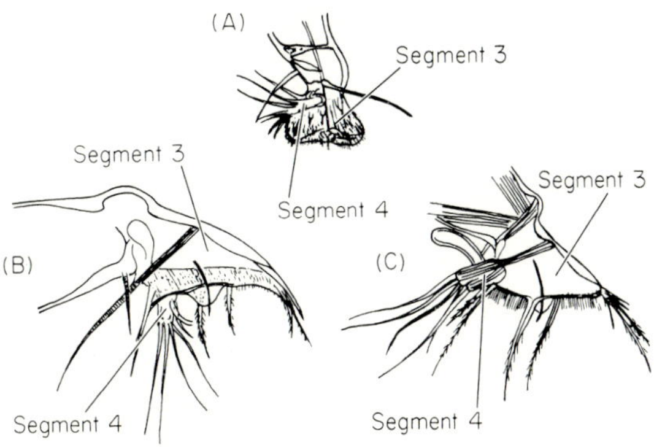

Fig. 2. Antennules of: A, *Tetraclita divisa;* B, *Scalpellum gibberum;* C, *Scalpellum ventricosum.* After Nilsson-Cantell, 1921.

the ducts from the cement glands open via pores on the disc. A number of setae—in external appearance perhaps characteristic of a given genus (Fig. 2)—arise laterally from the short and laterally placed terminal segment of the antennule.

The intrinsic musculature of the antennules is complex and has not been described; there are ten pairs of associated extrinsic muscles, which are shown in Fig. 1.

The cement glands, kidney-shaped bodies, are situated somewhat posterior to the large, prominent, lateral compound eye. Each gland consists of a compact mass of elongated secretory cells, each of which communicates with a collecting duct running dorsally on the median face of each gland; each glandular cell is some 70 μ in length, with a nucleus 8–10 μ in diameter, and with cytoplasm containing secretory material which remains as a reticulum with granules after fixation and staining. The cement duct, apparently lined with chitin, leads to an oval sac whose wall has an outer layer of circular muscle. This sac tapers to join the terminal portion of the duct which enters the base of the antennule and runs along it to the adhesive disc on which it opens by a number of fine pores (see p. 93).

It should perhaps be mentioned that the "gut-formed" masses into which Darwin traced the cement ducts are almost certainly the cement glands, but there is no evidence to support his suggestion that they give rise to ovarian tissue. Doochin's account (1951) of the cement glands as being present within the antennules appears to be erroneous as does his suggestion that Darwin's cement glands are "shell" glands. Although Bernard and Lane (1961, 1962) recognized the cement glands, their failure to find the cement ducts led them to the erroneous conclusion that the cement was extruded down the outside of the antennules. It would appear that as a result of neural mechanisms acting at settlement and in response to stimulii arising from contact with the substratum the cement glands are triggered to release their contents into the muscular sacs from which they are expelled by peristalic contractions to the exterior.

Fine Structure of the Antennule

It has been clear for some time that further advances in the interpretation of the behavior of the cyprid during ambulation and at settlement depend upon increased knowledge of the fine structure of the antennule and ultimately of the nerve physiology of the receptor and effector elements involved. Nott (1969) and Nott and Foster (1969) have recently provided the first of these requirements, namely,

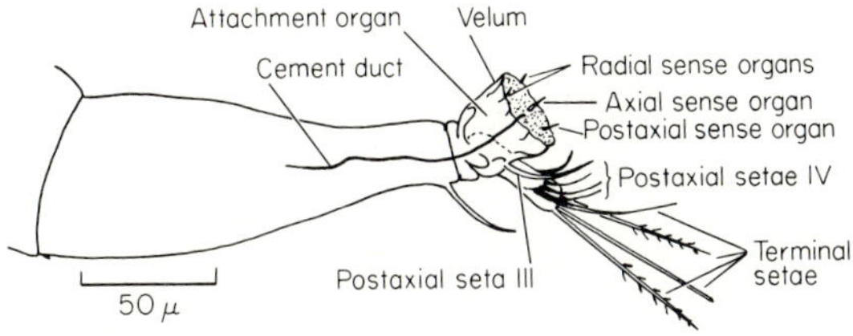

Fig. 3. Antennule of *Balanus balanoides*. From Nott and Foster, 1969.

a detailed account of the structure based on an electron microscope study. Their account, which refers to *Balanus balanoides* (Figs. 3-6) and concentrates on the bell-shaped third segment, will be closely followed (see footnote, p. 90).

Both the surface and margin of the disc are covered by cuticular villi among which a complex pattern of narrow slits leads into radial cement ducts, which arise from the axial cement duct. Secretions from

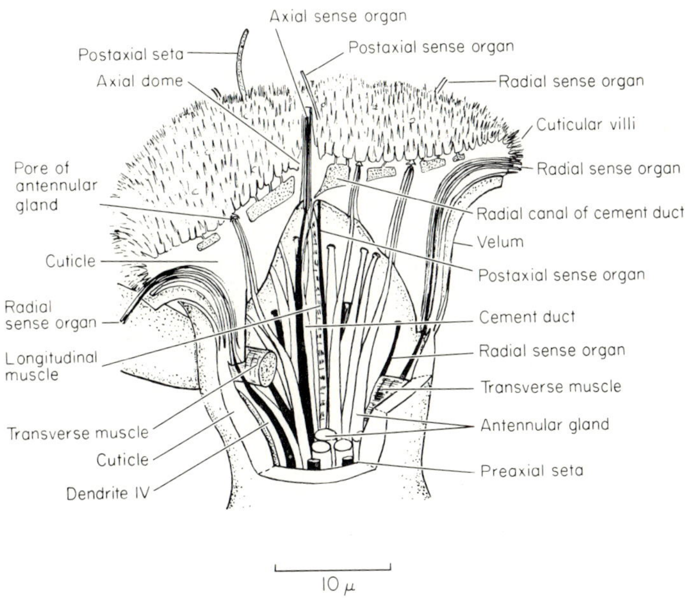

Fig. 4. Fine structure of the antennule of *Balanus balanoides*. Diagram resulting from interpretation of results of electron microscope studies. From Nott and Foster, 1969.

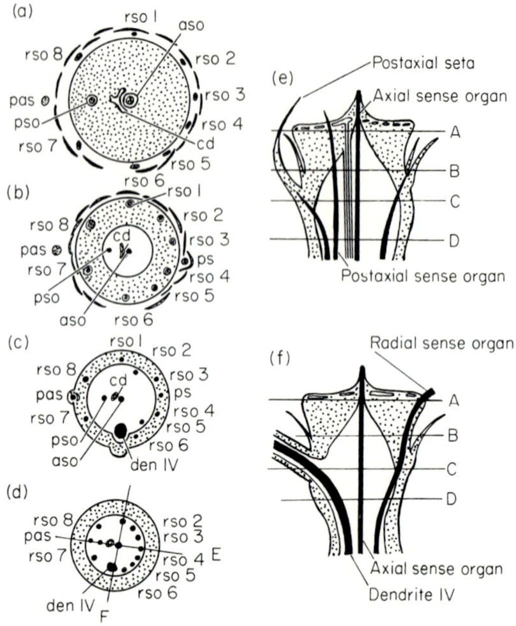

Fig. 5. Arrangement of sense organs: schematic transverse and longitudinal sections, aso, axial organ; cd, cement duct; pas, postaxial seta; ps, preaxial seta; pso, postaxial sense organ; rso, radial sense organ; (a)–(d) transverse sections at levels A–D; (e) and (f) sections in plants E and F in (d). From Nott and Foster, 1969.

the antennulary glands, lying in the second and third segments of the antennule, reach the underside of the disc via pores in its cuticle. The axial and post axial sense organs are tubular hairs mounted on a dome of cuticle, the former projecting from the center of the disc, the latter through an excentrically placed pore toward the postaxial side. Other hairs emerge radially from openings in the cuticle between the proximal margin of the disc and thickened velum which is made up of many sheets of cuticle attached midway along the side of the bell and ensheathing its distal region. An annulated postaxial seta arises midway along the third segment, and a smaller preaxial seta is hidden under the velum.

The fourth segment of the antennule which, as described above arises laterally on the third segment, bears four terminal setae; two of these bear setules, one is short and without setules, and the fourth, also without setules, is a long rodlike structure. In addition, the postaxial protuberance on the fourth segment bears four annulated setae curved towards the attachment disc. There are two muscles within the segment. Most of the structures originate proximally to the third segment and terminate at the disc. The axial sense organ, cement duct, and longitudinal muscle traverse the length of the third segment along the axis of the bell and are surrounded by glands and the ring of sense organs. The transverse muscle crosses the third segment at right angles to the axis to be inserted into the base of the fourth segment.

Nerve cells proximal to the third segment give rise to the sensory components of axial sense organ which terminates as an open cuticular hair. In the proximal region of the third segment the axial sense organ has five dendrites enclosed within longitudinal invaginations of a single sheath cell, so that each dendrite lies within a double membrane, the outer component being either continuous with the plasma membrane of the sheath cell or the membrane of another dendrite, so that in this region there is direct contact between dendrites. There are ciliary roots and bases midway along the attachment organ, and sup-

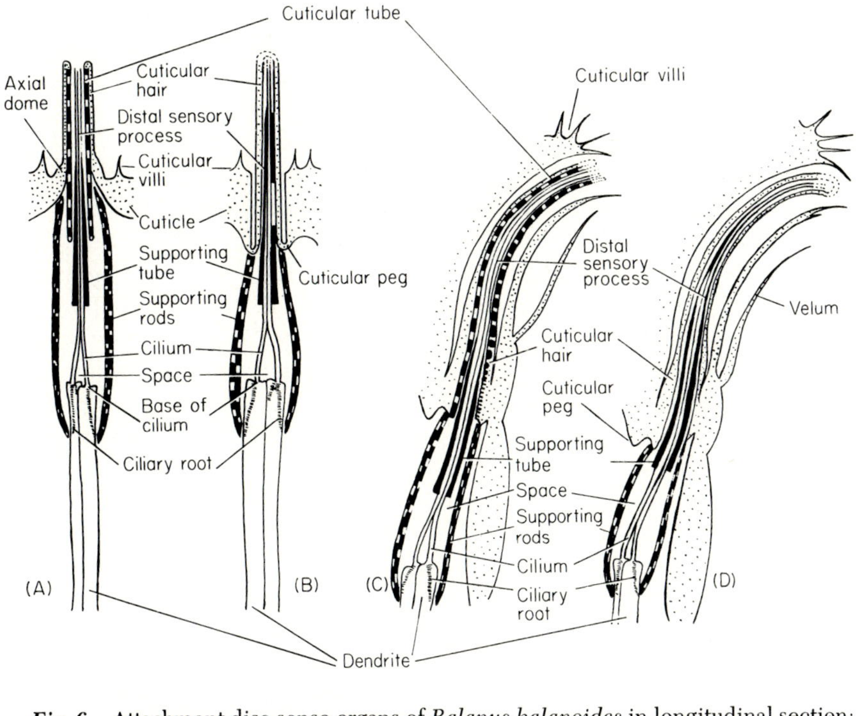

Fig. 6. Attachment disc sense organs of *Balanus balanoides* in longitudinal section; number of dendrites reduced to two in all cases; sheath cells omitted; schematic; (A) axial sense organ; (B) postaxial sense organ; (C) radial sense organ with internal tube; (D) radial sense organ without internal tube. From Nott and Foster, 1969.

porting rods in the sheath cell. Distal to this region five cilia are freely suspended in an extracellular space surrounded by longitudinal folds of the sheath cell; supporting rods are present in the folds. Immediately below the cuticle of the disc the cilia become five tightly packed distal sensory processes, four of which are similar to the cilia in cross section while the fifth is considerably dilated and irregular in outline. Folds of the sheath cell with the supporting rods continue to ensheath the tube in a longitudinal invagination of the cell. At the level of the inner margin of the cuticle of the disc, the supporting tube distally enters a cuticular tube extending the length of the axial hair, after which the wall of the supporting tube becomes gradually thinner; at the level of the surface of the attachment disc four distal sensory processes are enclosed directly within the tube. The single process with an irregular outline terminates within the supporting tube, but the other four processes, with limiting membranes and microtubules, continue as such towards the apex of the axial dome. The supporting rods of the sheath cell terminate on the cuticle of the disc but the sheath cell itself continues into the dome where it only partially ensheaths the cuticular tube which lies in a longitudinal groove on one side of the cell. The tube appears to be open, the cuticular tube folding outwards at the tip of the hair to become continuous with the exterior cuticle.

In the proximal region of the third segment, two dendrites in contact with each other and enclosed within a longitudinal invagination of a sheath cell make up the postaxial sense organ. Midway along the segment each sensory process gives a cilium which lies in an extracellular space, in turn bounded by longitudinal folds of the sheath cell and containing numerous supporting rods. Immediately below the cuticle of the disc, the form of the cilia changes and becomes a distal sensory process. The supporting tube passes through a peg of cuticle projecting into the lumen of the segment from the disc and continues into the lumen of a cuticular hair; the supporting rods of the sheath appear to be attached to the projection. Except at its base, the sensory hair is unattached, and it passes via a canal through the cuticle of the disc. At the base of the hair one of the sensory processes terminates while the other continues distally, terminating only as the hair emerges from the disc. Since the hair does not contain an inner cuticular tube, Nott and Foster believe it to be closed.

There are two kinds of radial sense organs, namely, radial sensory hairs with and without tubes. They all terminate as cuticular hairs emerging from the disc. All have certain features in common. In the proximal region of the third segment these sense organs are arranged in a circle midway between the axis and the wall. Each organ consists of a group of dendrites within longitudinal invaginations of a sheath cell but in contact with one another in the region where the ciliary roots develop; numerous supporting rods are present, terminating proximally in intimate contact with the dendrites and distally attached to the cuticle. The ciliary roots diverge from the axis of the sense organ and come to lie against the plasma membrane of the dendrites adjacent to the supporting rods. Midway along the segment the dendrites give way to cilia, suspended in an extracellular space bounded by the inner fold of the sheath cell. The orientation of these radial sense organs relative to the axis is such that they remain in a circle around the central structures but come to lie adjacent to the cuticular wall of the segment when distal to the transverse muscle. In this region the processes continue obliquely into the lateral wall and enter a cuticular peg projecting into the lumen of the segment and thence through a cuticular canal into a cuticular hair. These hairs continue in an oblique direction through canals in the wall, turn distally through 90 degrees, to emerge at the proximal side of the attachment disc, and directed outwards from the axis of the segment.

The two radial sensory hairs with internal tubes are similar to the axial sense organ in possessing five sensory processes and a cuticular tube within the hair. The five cilia become distal processes ensheathed within a supporting tube, which enters a cuticular tube, and these together penetrate the cuticle of the organ and enter the base of the hair. The supporting tube terminates, leaving the distal process enclosed by the cuticular tube. The tube is always excentric within the folded outline of the hair and in close contact with the exterior cuticle of the shaft on that side away from the longitudinal axis. Within the base of the hair the number of distal processes is reduced, and where the hair emerges among the lateral cuticular villi the thick inner tube contains two or three sensory processes with microtubules. Nott and Foster believe these organs to be open. The radial sensory hairs without an internal tube are more numerous and in structure resemble the postaxial sense organ, i.e., there is no cuticular tube within the sense organ and only two or three sensory processes are present. Initially, the cilia, arising from the dendrites, lie within a large extracellular space formed by a longitudinal fold in the sheath cell, but distally they are tightly enclosed within the sheath cell. Near the entry of the sense organ into the cuticle the cilia change to distal sensory processes within a supporting tube which penetrates the cuticle and enters the base of the hair. The number of sensory processes is reduced distally and near the margin of the disc the tube is empty.

The radial hair on the median side of the disc, with four cilia within the lumen of the segment, is somewhat different from the others. In addition to these sensory organs—all ultimately associated with the disc—the more proximal part of the third segment bears pre- and post-axial setae which pass obliquely through the cuticle. That on the post-axial side is much stouter than the radial hairs and has a supporting tube with sensory processes, the tube penetrating the base of the hair which emerges from the cuticle between the velum and the body of the third segment.

Numerous glands, the antennular glands—quite distinct from the cement glands—are present within the proximal part of the third segment of the antennule. Proximally, they are arranged around the axial structures but toward the disc are nearer the walls. Each gland is surrounded by a longitudinal fold of a sheath cell distally containing numerous supporting rods and with longitudinally orientated microtubules packed marginally between the secretion and the limiting membrane. The secretion itself is present in vesicles bounded by a membrane. Distally, each gland is enclosed in a double membrane derived from the sheath cell, and an extension of the inner fold of the sheath cell containing supporting rods is also present in the tube within which the gland and sheath cell extension terminate. Beyond this termination the tube becomes a flattened duct passing through a wide canal in the cuticle of the disc; a process from the sheath cell extends into the cuticular canal with the duct. Distally the canal is terminated by a thin layer of cuticle which is folded into a cone sunk beneath the surface of the disc; there is a pore at the apex of the cone. The ducts from the outer circle of glands pass obliquely through the cuticle and the pores are present near the margin of the disc, while the ducts from the inner circle of glands pass through the cuticle parallel to the axis with pores around the axial sense organ.

Nott and Foster have also described the fine structure of the ducts of cement glands in their passage through the third segment of the antennule. Proximally, the duct is circular in cross section, but beyond the transverse muscle this gives way to a crescent shaped section, which is lost as the duct reaches the cuticle of the disc; here the cuticular wall develops folds which eventually extend as radial canals passing through the cuticle of the disc and branching almost to reach the margin. They open on the surface by numerous slits.

The fourth segment is almost completely filled by the dendritic processes of nerve cells, showing at various levels differences in form—dendrite, cilium, distal sensory process—associated with sensory structures. Groups of dendrites and their sensory processes are surrounded by a fold of a sheath cell and appear to be associated with setae. The four curved setae on the postaxial side have an annulated appearance resulting from thickening bands beneath the cuticle of the hair shaft.

It is clear that these sensory elements show many features in common with similar structures previously described from other arthropods, even though previous accounts largely refer to adult and not larval stages. All the sense organs of the disc are comparable with the chordotonal scolpodia described by Howse (1968) in the termite, the supporting tube being analogous to the scolopale cap.

III. Reactions at Settlement

The duration of the planktonic phase of the cyprid under natural conditions is not precisely known for any species. Recently Walley (1969) has stressed that the process of metamorphosis should not be considered as changes taking place only at the cyprid-adult transformation but should rather be regarded as a series of progressive changes taking place throughout larval life and leading to the young adult. The organ systems of the cyprid gradually develop during the nauplii stages IV–VI, not simultaneously, but at a time closely related to that at which a given organ must begin to function. The most obvious and most rapid changes do, however, take place at the nauplius VI–cyprid moult and after settlement. Walley considers that the rates of change allow the metamorphosis to be rather considered as a two-stage process, namely, nauplius to cyprid and cyprid to adult. It is in the cypris stage that the complex sensory organs, including the paired eyes and the antennules with their adhesive apparatus, which are particularly adapted to function during settlement, become fully developed. It seems certain that since the cypris antennules are not fully differentiated for some time after the nauplius VI–cypris moult they will not be able to act as sensory organs relative to a substratum for some time after this moult. There have been no successful attempts to induce metamorphosis by means other than contact with a solid surface and it seems that the stimulation resulting from such a contact is essential to initiate the processes that lead to the reorganization accompanying metamorphosis, although once initiated it may proceed further. It is generally thought that any surface on which settlement takes place is encountered during the course of random swimming movements within a static or moving body of water; as a result of the color- or light-reflecting properties of underwater surfaces, there may, however, be directed movements of a cyprid toward them. Simple contact may not be enough to initiate settlement; a surface is "explored" (see below) and during this exploration the cyprid shows a distinct and characteristic behavior pattern. The behavior in relation to the various properties of solid surfaces has rarely been determined by direct observations but is usually inferred from the effects of variously modified surfaces on subsequent settlement patterns in which only the final result of the behavior is seen. It has been repeatedly shown, for example, that surfaces differing in texture or contour collect very different numbers of cyprids when exposed to the same planktonic populations. Experiments of this kind show, for example, that rough surfaces collect many more cyprids than smooth ones. In general, it cannot be assumed that more cyprids arrive on the rough surfaces; arrivals on the two kinds of surface must be assumed to be similar and a greater proportion must, therefore, leave the smoother surface.

A. *The Ambulatory Phase*

The cyprid, even of an intertidal species, begins its ambulatory phase and settles under water. According to Knight-Jones and Crisp (1953), cyprids, when exposed to air, release their hold and are easily washed off. The animal moves forward in the following way (Pyefinch, 1948; Knight-Jones and Crisp, 1953; Barnes, 1955): After one antennule has been firmly attached, its antennulary muscles contract and the body, attached only by this antennule, is, therefore, drawn forward. Both Darwin (1854) and Claus (1869) earlier drew attention to the large size of the antennulary muscles. The free antennule is next firmly attached, the other is released, and the sequence repeated. Even in the early phases of ambulation, "testing" movements are made (Barnes, 1955). The free antennule is repeatedly put down over a small area around the anterior of the larva, rather like a horse pawing the ground, after which temporary attachment followed by forward movement may take place. From time to time the animal may "rest" with one antennule temporarily attached, then lift the body clear of the surface and make vigorous "swimming" movements with the thoracic appendages similar to those of the planktonic phase (Barnes, 1955; Bernard and Lane, 1962). The impression is gained that the animal, by these tugging and swimming movements, either together or separately, is trying to release the antennule and thereby test the adhesive properties of the surface. At the end of this ambulatory phase the cyprid comes to rest with both antennules on the surface and rotatory movements are made with the point of attachment as pivot (Doochin, 1951). These movements, which gradually decrease in frequency and extent, may be accompanied by further activity of the thoracic appendages similar to that during resting periods within the ambulatory phase. In the later stages, such movements are interpreted by Knight-Jones and Crisp (1953) as serving to free the substratum from silt or loose particles. The cyprid then pulls itself down on its antennules.

It seems certain that the attachment during the ambulatory phase is physically a quite different phenomenon from that at fixation. It has sometimes been assumed that adhesion during exploration is by suction but, as Nott and Foster (1969) have pointed out, the fine details of the structure of the disc do not indicate that it would be able to function as a suction pad; there is no efficient seal around its edge, the velum being unlikely to function as such since it is discontinuous around its edge, while the cuticular villi are disadvantageous to such a function. Unless the spaces in the velum and between the marginal villi were filled with liquid the disc would be difficult to seal, and although the annular gland fluid might provide such material it does not discharge at the periphery. Further, when the larvae is subjected to a force such as experienced in a water current which leads to stronger yet not permanent adhesion, a mechanism is required to reduce the pressure between the disc and the substratum, the center being pulled away while the periphery remains attached if suction is the relevant adhesive force. As Nott and Foster state, there appears to be no obvious mechanism for this since, although the peripheral cuticle of the disc is attached, so is the central region and the only muscle which appears in a position suitable for such a lift is very slender and terminates not on the disc itself but at the cement duct. Furthermore, the third antennal segment has no large hemocoele into which fluid could be drawn. Nott and Foster consider that the disc must present a true adhesive surface and they consider the secretion of the antennal glands may provide the necessary adhesive substance. Since the cyprid "walks," putting down each antennule one after the other, there must be some coordinating mechanism related to detachment. Temporary adhesion by means of a temporary secretion does not seem, on the other hand, to be easily capable of explaining the greater strength of adhesion under the action of a shearing force such as a current. It is unlikely that any stimulated increase in the secretion of material by the current would lead to better adhesion, and it also seems unlikely, although not impossible, that the efficiency of the adhesive material is a characteristic of the species and its normal habitat as is adhesion under current shear. Such stimulation could, however, lead to an inhibition of a nervous mechanism by which the detachment part of the cycle is triggered.

The part played by any particular aspect of this behavior pattern in maintaining the cyprid on a solid surface and in leading to ultimate settlement is not known, nor have we any information regarding possible metabolic or physiological changes initiated by ambulation. The fact that cyprids will not settle on a relatively unstable fibrous surface of glass wool, whereas they will do so on a stable surface of similar material, was thought by Barnes and Powell (1950) to indicate that a certain reaction to the pressure of the antennule on the surface is essential if the ambulatory phase is to be maintained.

B. Effect of Water Currents

The first quantitative measurements on the effect of current speed on settlement were made by Smith (1946). For the most part, he used the rotating disc technique commonly employed in corrosion studies. He also made observations on settlement in glass tubes, in which a variation in current speed was obtained by the use of tubes of different diameters and a constant head of water. It is a simple matter to calculate the velocity at any point on the surface of a revolving disc, and this, the "nominal" velocity, was used by Smith although he was well aware of the difficulties involved in relating this to the true velocity of the water movements at its surface. His results gave a nominal limiting velocity of 0.9–1.3 knots (mean 1.1 knots) for the settlement of cyprids of *Balanus amphitrite niveus*; above this speed no settlement took place. A rather lower critical speed, 0.5–1.0 knots, was found for the same species in the experiments using tubes of different diameter; this second technique indicated that *B. improvisus* could settle at a somewhat greater, and *B. eburneus* only at somewhat lower, current velocity than *B. amphitrite niveus*.

As Crisp (1955) has pointed out, the important dynamic feature of water flow relative to settlement is likely to be the velocity gradient, or rate of shear, at the surface rather than the stream velocity within the body of the fluid; once within the boundary layer, the cyprid, small and with a density similar to that of the water, will, when not swimming, take up the velocity of the water. The velocity gradient at the boundary varies in the same sense as that in the stream but the relation between the two is dependent upon topography. Crisp examined cypris behavior under varying rates of shear using for the most part glass tubes varying from 2 to 10 mm in diameter; water was supplied from a constant head device, the quantity passing in unit time being determined by measurement of the effluent volume. Cyprids were introduced singly and their behavior, as they passed down the tube, observed. In contrast to Smith, who observed only ultimate settlement, Crisp regarded as "successful" any temporary attachment made by the cyprid as it passed between marks 1 meter apart on the tube. At low rates of flow, cyprids of *Balanus balanoides* did not swim but were rolled passively down and along the walls of the tube. Figure 7 shows the percentage of successes at different rates of shear for this species and for *Elminius modestus.* The highest percentage of attachment for *Balanus balanoides* takes place at rates of shear between 60 and 80 sec^{-1} (a gradient of 100 sec^{-1} corresponds to a water velocity of 1 m/sec at a distance of 1 cm from the surface under conditions of laminar flow); in this range the cyprids were stimulated to greater swimming activity and to move against the flow. This stimulation to greater activity supports the hypothesis of the maintenance of swimming in the free planktonic phase by proprioreceptive stimulation. Attachment was more difficult at rates of shear greater than 100 sec^{-1}; above a value of 400 sec^{-1} there was no attachment. A moderate

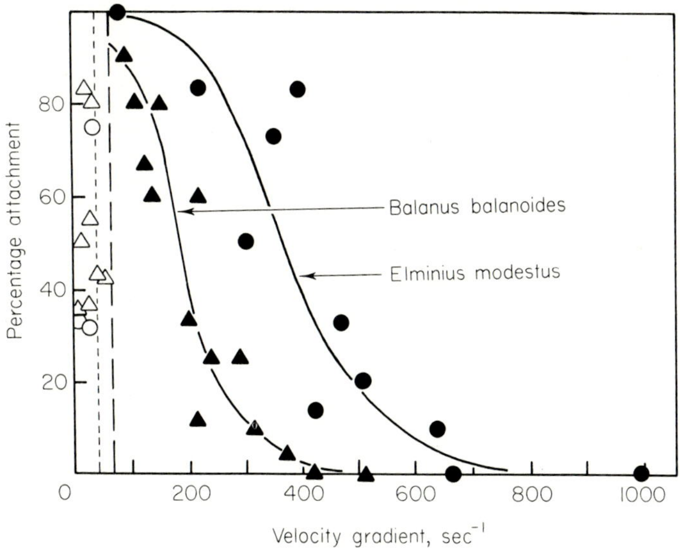

Fig. 7. Effect of velocity gradient on attachment of cyprids. Open symbols indicate little or no reaction to current; vertical dotted lines indicate minimum gradient to stimulate strong attachment reaction. After Crisp, 1955.

current not only stimulated attachment but also tended to keep a cyprid attached so that it could be disengaged only by reducing the current speed. Cyprids of *Elminius modestus* responded to rather smaller rates of shear with maximum attachment at values less than 50 sec^{-1}, and the passive behavior at low rates of shear was less marked. The attachment of *Elminius* cyprids at lower rates of shear is in accordance with an ecological difference, namely, that this species is largely restricted to quiet water situations, whereas *Balanus balanoides* is common in relatively exposed situations. Nevertheless, *Elminius* could attach at higher rates of shear (400–700 sec^{-1}) than *Balanus balanoides*; the passive behavior of the latter may have been due to their poorer condition for they will swim and, indeed, attach in "perfectly" quiet water. Crisp considered that the ability of cyprids of *Elminius* to settle at higher rates of shear than those of *Balanus balanoides* may be partly due to their smaller size; because of their size they will be closer to the surface at the moment of attachment and, with a given rate of shear, will be moving more slowly. On the other hand, conditions on the shore are complex, and factors not allowed for in these experiments may be important in the relative behavior of the two species. By appropriately shading the tube, one may induce the cyprids to walk to the illuminated side, either with or against the current. When they are walking in the direction of the current at high rates of shear, the posterior part of the body becomes lifted away from the surface. Crisp showed that, on a series of reasonable assumptions, the suctional force holding the cyprid to the surface (by the antennule) is adequate to maintain its position even when the current speed far exceeds the critical value for settlement, and calculations of the force required to remove the cyprid, once attached, agree with the values found in these experiments (see, however, p. 101). Even so, to walk as distinct from merely to hold on to a surface involves greater tensions in the muscles since the body is raised, for part of the time, from the surface and exposed to greater shearing forces; at rates of shear greater than 1000 sec^{-1} cyprids appeared to be unable to walk with the current and they either remained in one place or swam away. Because tubes of very different diameter were employed, there is no agreement between Crisp's results and those of Smith (experiments in tubes) when the mean flow velocity is used for comparison; recalculation of Smith's values in terms of rates of shear gives a rather lower value (100 sec^{-1}) for the maximum rate of shear at which he found settlement could take place. This discrepancy may be due in part to the fact that different species were being used, and in part to the fact that Smith observed only complete attachment leading to adult settlement.

The agreement between the results of the tube experiments and those of the rotating disc are difficult to explain. Values for the rate of shear at the surface of a spinning disc can, because of the complex hydrodynamic conditions, be only approximate, but even so the calculated critical values which are of the order of 1000–2500 sec^{-1} are widely different from those of the tube experiments.* Crisp points out that, relative to cyprid settlement, conditions on a revolving disc differ from those in a tube in one important respect, namely, that in the former there is a gradient of shear over the surface, a gradient which rises sharply from zero at the center to reach a maximum value at the edge. Attachment nearer the center under relatively low rates of shear may, therefore, be followed by migration outward across this gradient. On this view these critical values for the disc experiments are an expression of the maximum gradients across which a cyprid will walk. There is, however, little evidence that cyprids do migrate in this way, for in the ambulatory phase they seem relatively insensitive to water currents.

Crisp calculated that maximum attachment took place when the current speed at half the height of the cyprid from the surface was equal to the maximum swimming velocity (about 4.8 cm/sec for *Balanus balanoides*) and considered, therefore, that the cyprids attach when their movement against the stream is capable of bringing them to rest with respect to the surface. Although relative motion of the cyprid with respect to the surface is clearly important, this hardly seems a complete analysis. In the first place this would define the optimal speed in relation only to the size of the cyprid and to its swimming speed without reference to such species-specific characters as are known to exist. The optimal velocity for maximum attachment may well be the net result of a balance between the speed of swimming of the cyprid relative to the surface and the optimal drag necessary to initiate the adhesive reaction and inhibit swimming. The greater ease with which this inhibition takes place when the animal is swimming is perhaps a function of its generally greater nervous activity. It would perhaps help to resolve some of the problems associated with the mechanisms of attachment in currents if data were available on the relation between critical rate of shear and surface roughness.

*One criticism may perhaps be made of some tube experiments. In order to obtain a wide range of rates of shear under laboratory conditions where the head and total volume of water available at the head is limited, it is necessary to use tubes of small bore. Crisp used tubes from 1 to 3.2 mm radius, and the diameter of the smallest of these is only twice the length of some cyprids.

C. Gregarious Behavior during Settlement

Barnacles of many species tend to settle upon each other, forming large and even massive clumps. In the intertidal zone this may often be only an expression of the limited space available for a heavy spatfall and under these conditions it often leads to self-destruction of the population and can have no survival value (Barnes and Powell, 1950). Aggregations of sublittoral barnacles are, however, not uncommon and indeed characterize some species. They have been particularly noted in *Balanus balanus* and *B. hammeri* and perhaps are most marked in *B. nubilus* because of the large size of the individuals, clumps of which may be 1–2 ft high and weigh several pounds.* It may be that such aggregations again result from the effect of space limitation, adult barnacles being largely the only suitable substratum in that particular place for successive spatfalls. On the other hand, it is possible that the surface of the same or another species of barnacle is more attractive to cyprids. That this is the case that was first put forward by Knight-Jones and his colleagues in a number of papers relating to what they term gregarious behavior. The term gregarious is more commonly applied to motile animals to indicate some intraspecific attraction holding them together in groups and in its most complicated manifestation giving rise to higher social orders such as are found in the insects, with their complex degree of organization of the society and impressive specialization of individuals within the community. Gregarious behavior in cirripedes is at a low level of coordination.

Knight-Jones and Stevenson (1950) exposed small panels to a settlement of *Elminius modestus*. They found that more spat settled on those panels initially bearing barnacles than on the bare panels, the mean ratio for a series of experiments being about 3:1. An alternative experimental approach consisted of mounting a slide, bare or with barnacles, at the center of a panel and counting the number that subsequently settled at various distances from this central slide. Settlement was consistently higher on the slide bearing barnacles and its surroundings. Moreover, neglecting those settled due to rugophily at the junction of the central slide and its surroundings, the authors concluded that settlement fell off with increasing distance from the central slide when it bore barnacles, but not when it was originally bare. They drew attention to the fact that, unlike the previous experiments,

*Aggregates are common in deep sea pedunculates and seem to reach an extreme form in *Scalpellum sociabile* from the Bali Straits (Annandale, 1905).

where settlement in groups may result from contact with cyprids and other adults, this experiment suggests action at a distance; settlement in response to a water-borne substance whose effect would be expected to decrease with increasing distance from its source as a result of dilution was, therefore, postulated. They also point to field observations in which settlement was found to be heaviest on those parts of the shore where adult barnacles were abundant, a phenomenon which they attributed to the stimulatory effect of the adjacent adults; a sensory basis for the attraction was again postulated. Such field evidence is, perhaps, less conclusive than the experimental work in supporting any theory of interaction between adults and cyprids. That the environment was favorable to barnacles is evident from their presence in large numbers and the favorable nature may or may not have been acting at the settling phase. The greater settlement could be due, not to the presence of a population of adult barnacles per se, but to those factors of the environment that had originally favored the establishment (and possibly continuance) of that population. The greater part of the earlier work of Knight-Jones and all subsequent work has strongly indicated that gregarious behavior is dependent upon contact with a solid. The results of these experiments in which action at a distance was implicated remain largely unexplained (see below). Knight-Jones and Crisp (1953) have pointed out that a cyprid, once it has contacted an adult barnacle, will more readily settle on a bare surface; they suggested that a cyprid can detect a quinone-tanned protein in the epicuticle, since only those reagents, such as sodium hypochlorite and strong acids, which attack this material were found to be capable of destroying the attractive nature of adult bases. The results so far discussed have been largely obtained by considering the settlement on panels exposed under more or less natural conditions to populations of cyprids. Knight-Jones (1953a) has compared under laboratory conditions settlement in the presence and absence of barnacle bases. He found that the presence of *Mytilus* shells bearing barnacles induced settlement and that species could distinguish their own adults from those of other species. Further, it was shown that cyprids that had encountered barnacles in the experimental dishes received what Knight-Jones terms a sensitizing stimulus, so that subsequently they settled more readily on glass or other relatively unfavorable surfaces. Stimulus to settlement was provided by newly settled cyprids as well as by the presence of adults. Furthermore, when fragments from various parts of the adult were placed upon surfaces which had never borne barnacles, they made these surfaces somewhat more attractive for settlement. As a result of more detailed experiments on the nature of the active substance, the hypothesis that it is a quinone-tanned protein was reaffirmed, but the previous suggestion that water-borne material and chemoreception are involved was abandoned. Contact was considered to be essential. Knight-Jones also stressed that, within any population, the level of discrimination during settlement may well vary and, like that of the serpulid *Spirorbis* (Knight-Jones, 1953b), may be slightly lowered after prolongation of the free-living stage. More recently (Knight-Jones, 1955), evidence has been presented that the induced settling reactions are to some extent species specific. *B. balanoides* was found to settle more readily on stones bearing this species than on those bearing adult *Chthamalus stellatus* or *Verruca stroemia.*

Since, with the single exception already noted, it was evident from Knight-Jones' experiments (1953a) that cyprids respond only to contact with bases and that the latter are effective even after various chemical treatments, Crisp and Meadows (1962, 1963) have investigated the possibility that an adsorbed layer is involved since they considered that the alternative, namely, that a stimulatory substance is released by enzyme action on contact, is more difficult to reconcile with specific reaction at a molecular level (see, however, below). They have shown that panels treated with barnacle homogenates are more attractive to cyprids and that the settling factor is a property of the extract itself and is not developed by bacterial action during its exposure to seawater. Active material could not be prepared from algae but was present in extracts of two sponges, of the fish *Blennius pholis*, and particularly in homogenates of all the arthropods tested. Using *Carcinus maenas* it was shown that activity is restricted to the carapace and that newly moulted animals give more active extracts than those with a fully hardened exoskeleton. These facts together with the chemical properties of the settling factor—relative heat stability, no diffusion on dialysis, and resistance to hydrogen peroxide and formalin—led them to identify the factor with the protein component of the arthropod exoskeleton. They considered that the factor is effective as an adsorbed layer and not as a result of diffusion from the surface. Although after immersion in a solution of the settling factor when cyprids showed a greater tendency to explore inert surfaces, there was no increased settlement neither then nor when the cyprid was allowed to explore a surface contained in a solution of the settling factor; no preference was shown for panels surrounded by an extract and separated from a control by a density gradient. Further, cyprids discriminate between treated and untreated surfaces when both are presented in seawater containing the active material in solution. The effect of ammonium sulphate, which increases the tendency to form adsorbed layers, is to favor settlement; curiously no progressive increase in settlement was found with increasing salt concentration. The effect of formaldehyde on either the solution of active material or on the panels after treatment with it was to increase settlement; this is interpreted either as the result of increased protein adsorption on the panels or as modification of the settling factor to give a more effective molecular structure. By exposing cyprids of various ages to test panels, Crisp and Meadows (1963) found an increase in the initial settling rate with cypris age, suggesting that, with respect to the settlement reactions, there are physiological changes with time. They did not find any change in discrimination with age; treated surfaces were always selected. This contrasts with the behavior of *Spirorbis*, in which the threshold of surface discrimination falls with age (Knight-Jones, 1953b).

Chemical attraction by water-borne substances is well known, but in the case of a planktonic larva which has to find a surface for a sedentary adult two features make diffusion and chemotaxis less effective. The smaller the animal and the smaller the distance separating the receptor organs the more difficult it is for it to respond to some forms of chemotaxis. Under these conditions it is not surprising to find that the larvae of many sedentary organisms encounter surfaces during random movements and only after being encountered do the surfaces exert retaining properties, specific or otherwise. Such properties could most easily be provided by an adsorbed layer.

This property of the active substance when present on surfaces cannot be attributed to an increased concentration within the adsorbed layer; the molecules in the layer are no more accessible than those in solution. The reason why the cyprid responds similarly to adsorbed layers deposited from solution and to "insoluble" cuticular material may lie in the fact that the protein is adsorbed on the surface by its polar groups and may present a similar molecular configuration to the medium as the same protein bound in the epicuticle.

Three structures in the cyprid antennule would seem to be comparable with the known chemoreceptors in other arthropods, namely, the axial sense organ and two of the radial sense organs (Nott and Foster, 1969). They possess an internal cuticular tube surrounding the distal sensory process within the hair which in the axial sense organ has been shown to be open at the tip and which from a comparison of the fine structure probably is so in the two radial sense organs. One of the processes in the axial sense organ and some in the radials terminate within the supporting tube so that the organs may function as both as mechano- and chemosensory organs, a situation known in other animals; the tube could respond to a distortion of the attachment organs and in this way would function as mechanoreceptors. Nott and Foster (1969) have considered gregarious behavior in relation to their work on the fine structure of the adhesive organ. They consider that the recognition of specific compounds probably depends upon a normal chemosensory response to a stimulating substance by the sensory processes of some sensory cells and that a specific protein may not be recognized as such by a chemical tactile sense as suggested by Crisp and Meadows (1963); the protein of the attractive integument may be broken down by enzymes as originally suggested by Knight-Jones (1953a)—perhaps discharged from the antennular gland secretion and the products then detected by the chemosensory system. Chemoperception of amino acids is well documented. Such a mechanism has the advantage that it would not be so hindered by bacterial slimes (which soon cover any surfaces exposed in the sea) as would the tactile chemical sense proposed by Crisp and Meadows; it would explain why the cyprid does not respond to protein in solution, under which conditions the products of enzyme activity would be quickly dispersed from the sensory hairs; the greater concentration of products released from a monomolecular film under the adhesive disc might be sufficient to elicit a response from an adsorbed film but not from a volume of solution. In spite of this it must not be forgotten that although cyprids are attracted to protein bearing surfaces, they *will* attach to "clean" surfaces, particularly when discrimination is lowered as during a heavy settlement; as far as is known, no chemosensory sense is then involved.

References

Annandale, N. (1905). *Mem. Asiatic Soc. Bengal* **1**, No. 5, 73-84.
Barnes, H. (1955). *Vidensk. Medd. Dansk Naturh. Foren.* **117**, 341-348.
Barnes, H., and Powell, H. T. (1950). *J. Animal. Ecol.* **19**, 175-179.
Bernard, F. J., and Lane, C. E. (1961). *Biol. Bull.* **121**, 438-448.
Bernard, F. J., and Lane, C. E. (1962). *J. Morphol.* **110**, 19-40.
Bohart, R. M. (1929). *Am. Naturalist* **63**, 353-361.
Claus, C. (1869). *Naturw. Ges., Marburg Suppl.*, Part 5, 1-16.
Crisp, D. J. (1955). *J. Exptl. Biol.* **32**, 569-590.
Crisp, D. J., and Meadows, P. S. (1962). *Proc. Roy. Soc.* **B156**, 500-520.
Crisp, D. J., and Meadows, P. S. (1963). *Proc. Roy. Soc.* **B158**, 364-387.
Darwin, C. R. (1851). "A Monograph of the Subclass Cirripedia," Vol. 1, The Lepadidae. The Ray Society, London.
Darwin, C. R. (1854). "A Monograph of the Subclass Cirripedia," Vol. 2, The Balanidae. The Ray Soc., London.

Doochin, H. D. (1951). *Bull. Marine Sci. Gulf Caribbean* **1**, 15-39.
Hoek, P. P. C. (1884). *Rept. Challenger Exped., Zool.* **10**, No. 3, 1-47.
Hoffendahl, K. (1904). *Zool. Jahrb., Morphol.* **20**, 363-398.
Howse, P. E. (1968). *Symp. Zool. Soc. London* **23**, 167-198.
Knight-Jones, E. W. (1953a). *J. Exptl. Biol.* **30**, 584-598.
Knight-Jones, E. W. (1953b). *J. Marine Biol. Assoc. U.K.* **32**, 337-345.
Knight-Jones, E. W. (1955). *Nature* **174**, 266.
Knight-Jones, E. W., and Crisp, D. J. (1953). *Nature* **171**, 1109-1110.
Knight-Jones, E. W., and Stevenson, J. P. (1950). *J. Marine Biol. Assoc. U. K.* **29**, 281-297.
Nilsson-Cantell, C. A. (1921). *Zool. Bidr. Uppsala* **7**, 75-395.
Nott, J. A. (1969). *Marine Biol.* **2**, 248-251.
Nott, J. A., and Foster, B. A. (1969). *Phil. Trans. Roy. Soc.* **B**, 115-134.
Pyefinch, K. A. (1948). *J. Marine Biol. Assoc. U.K.* **27**, 464-503.
Smith, F. G. W. (1946). *Biol. Bull.* **90**, 51-70.
Visscher, J. P. (1928). *Biol. Bull.* **54**, 327-335.
Walley, L. J. (1969). *Phil. Trans. Roy. Soc.*, **B256**, 237-280.

CHAPTER 6

HISTOCHEMISTRY OF BARNACLE CYPRID ADHESIVE FORMATION

Robert E. Hillman
and
Paul F. Nace

I. Introduction

Many marine organisms secrete, under water, adhesive materials which fasten the organism firmly to a particular substrate. The barnacle's ability to attach itself under the most adverse conditions, such as a pounding surf or periods of drying when the out-going tide leaves the barnacles exposed, is one of the best known examples of this capability. The barnacle is able to cement itself to a variety of substances including various metals, wood, concrete, plastic, and other synthetic materials, and the resulting attachment usually outlives the barnacle. In fact, fossil barnacles have been found firmly adhered to other shells after thousands of years.

Initial attachment by the barnacle occurs only during a short and specialized phase of the organism's life cycle, when the free-swimming cyprid larva settles on a suitable surface and secretes a sticky adhesive which firmly binds the larva to the surface. The attached larva quickly metamorphoses to the young adult and begins the sedentary phase of its life cycle. As the barnacle grows it continues to secrete material capable of maintaining the barnacle in position for the remainder of its life.

Although the adhesive-secreting abilities of the barnacle have been known for hundreds of years there has, until recently, been very little work on the secretory processes in these organisms. This was due to the inability of researchers to obtain the critical life stages in sufficient numbers for experimental purposes. It is now possible, however, to culture the larval stages of the barnacle in the laboratory, making available the number of these organisms necessary for experimental studies. By inducing these stages to settle and fixing them with suitable histochemical fixatives at the peak of their secretory activity, one can make histochemical analyses of secretory areas and products. In this way precise biochemical information on the synthesis and composition of adhesives produced by the barnacle can be obtained.

In view of the need for such an adhesive and the availability of the various life cycle stages of the barnacle, the William F. Clapp Laboratories of Battelle Memorial Institute undertook research to provide information on the histochemistry and biochemistry of barnacle adhesives.

Early observations on barnacle settlement and attachment were made by Darwin (1854), Groom (1894), Visscher (1928, 1930), Bohart (1929), Herz (1933), and Clarke (1947). More complete studies of behavior and morphology of cyprid larvae during settlement, attachment, and metamorphosis were made by Doochin (1951), Crisp (1961), and Bernard and Lane (1962). These studies indicate that the barnacle goes through several relatively complex free-swimming larval stages prior to attachment and metamorphosis to the sedentary adult form. Usually, the fertilized egg develops within the female to a free-swimming larval stage, called a nauplius larva, before being discharged into the water. The nauplius undergoes a series of molts before reaching the cyprid stage of development, which is also a free-swimming stage. Generally the larva remains a cyprid for approximately 2–5 days before settling on and attaching to a particular surface.

Visscher (1928) has described the attachment behavior and notes that once the cyprid larva has settled on a solid surface its principal means of locomotion are the antennules, upon which, or by which, sticky droplets are secreted. Bernard and Lane (1962) confirm these observations and point out that at first there is no tendency for the antennules to adhere to the substrate by hairlike projections at their tips, but that eventually it becomes harder for the larva to withdraw its antennules from the substrate. Eventually the process of attachment is complete, and the animal can no longer detach itself. During this attachment period, droplets of a material which is insoluble in seawater can be seen on the surfaces of the antennules. This material was presumed to be the initial adhesive (Darwin, 1854; Visscher, 1928; Herz, 1933; Bernard and Lane, 1962; Kaufmann, 1965).

The histology of *Balanus amphitrite* has been described in some detail by Bernard and Lane (1962); however, they were unable to resolve certain questions regarding the actual secretion of the adhesive material. They point out, for example, that droplets of insoluble secretion can be seen on the antennule surfaces; however, the antennule of the attached larva does not seem to be in actual contact with the substrate, but rather it ends in a depression in a disc formed by the adhesive. In addition, there does not appear to be any adhesive in the antennule, nor are there ducts from what they presume to be the adhesive-producing gland (the postoptic organ) into the antennules. They state, furthermore, that the location of the supposed secretory gland precludes the probability that its secretions could enter any duct within the antenna, because the gland adjoins the antennule only at its basal segment which is covered by a continuous chitinous exoskeleton. The cement disc itself is laminar and would require successive additions of material after the initial formation of the plate. This would mean that if the adhesive came through ducts in the antennules, the tip embedded in the hardened disc would be plugged by the first secretions of cement, thereby preventing further additions.

Lindner (1968) was also unable to discern a duct from the cement gland to the tip of the antennule and hypothesized a suction cup effect for the initial attachment, rather than an adhesive. Rees (Crisp, 1969) has, however, been able to find the cement gland duct. Also, Nott (1969) has shown the cement duct openings on the tip of the antennule, with the scanning electron microscope.

Some discrepancy also exists concerning the secretion of the adhesive by the adult organism. Most investigators believe that the cyprid cement gland develops to form the adult gland. The location of these glands, however, is in question. Arvy, Lacombe, and Shimony (1968) and Arvy and Lacombe (1968) believe that the adhesive is secreted by unicellular cement glandules dispersed among the ovarian follicles while Lindner (1968) feels that the glands are not located in the ovary at all, but rather higher up in the body of the adult organism.

These differences in opinion indicate that there is still much to learn concerning the formation and secretion of the adhesive and its chemical composition. The work described below is an attempt to provide information on the chemical composition of the cyprid adhesive, since most people agree upon the location of the ultimate secretory product. The approach used is a histochemical one, in order to avoid the difficulties of collecting for analysis pure material in the extremely limited amounts offered by the cyprid larva.

II. Materials and Methods

Laboratory reared cyprids of the barnacle *Balanus eburneus* were induced to settle on and attach to preconditioned strips of balsa wood. The attached larvae were then fixed either in Bouin's solution, which aided in decalcifying as well as fixing, Carnoy's fixative for basic proteins, Zenker's solution, 10% formalin in seawater, formalin-sublimate acetic acid (Movat, 1955), 10% formalin with calcium acetate and cetylpyridinium chloride added for carbohydrate containing tissues, or in McManus' solution for lipid histochemistry.

In some cases, the larvae were allowed to metamorphose to the young adult stage before fixation in order that the cyprid's adhesive might be compared with that produced by the young adult.

Following fixation for 48 hours, the cyprids were dehydrated through a graded series of alcohols and xylene, and imbedded in paraffin for sectioning at 5 μ.

For general histological purposes, the barnacle cyprid's tissues were stained with iron hematoxylin and eosin, Mallory's triple stain, a quad-type stain (Wismar, 1966), and a trichrome stain (Green and Wood, 1959).

Carbohydrates, particularly mucous materials, were studied by means of the Periodic acid–Schiff (PAS) reaction with diastase controls, alcian blue–PAS sequence (Mowry and Winkler, 1956; Spicer, 1960) aldehyde fuchsin–alcian blue (Spicer and Meyer, 1960), high iron diamine/alcian blue (Spicer, 1965) and toluidine blue O (Michaelis, 1947).

Lipids were localized by staining with Sudan black B following fixation with McManus' fixative.

Proteins containing phenolic amino acids were stained by means of the Millon reaction (Bensley and Gersh, 1933; Pearse, 1960). Baker's modification of the Sakaguchi reaction (Pearse, 1960) was used to localize proteins containing arginine. Gurr's picro-ponceau reaction (Humason, 1967) was used specifically for collagen or collagenlike proteins. Basic proteins were stained with Biebrich scarlet following Carnoy fixation (Spicer, 1963). A diazosulfanilic acid–azure A sequence (Lillie *et al.*, 1966) and the bromsulfalein technique of Silverman and Glick (1966) were used as general protein stains.

Possible neurosecretory substances were looked for with Bargmann's chrome hematoxylin stain (Pearse, 1960).

A general staining was accomplished with Gomori's aldehyde fuchsin (Pearse, 1960) and with an azo blue-Congo red combination in which the tissue was stained for 15 minutes in a 0.1% aqueous solution of azo blue followed by 15 minutes in a 0.1% aqueous solution of Congo red.

III. Results

The results of the histochemical reactions on the sections of attached *B. eburneus* cyprids are described in Table I.

The adhesive material, shown in Fig. 1 at the top of the antennules, did not exhibit any noticeable reactions for carbohydrate-identifying materials and only a trace of positive reaction for the lipophilic Sudan black B. On the other hand, most of the general reactions for proteins, some exhibited by histological stains rather than histochemical reagents, gave strong positive results. The specific amino acid identifying reagents exhibited principally negative results. For example, the adhesive was eosinophilic with hematoxylin and eosin, and stained

TABLE I
SUMMARY OF HISTOCHEMICAL REACTIONS ON SECTIONS OF ATTACHED CYPRIDS OF THE BARNACLE *B. eburneus*

Reaction	Result
Hematoxylin and eosin	Reddish-pink
Mallory's triple stain	Red-orange
Quad-type stain	Bright red
Trichrome stain	Yellow-green
PAS	–
PAS and diastase	–
Alcian blue-PAS	–
Aldehyde fuchsin-alcian blue	–
Toluidine blue O	–
Sudan black B	Trace black
Millon	–
Sakaguchi	–
Picro-Ponceau	Red
Biebrich scarlet	Bright red
Diazosulfanilic acid-azure A	Pale yellow-green
Bromsulfalein	Purple
Bargmann's chrome hematoxylin	Phloxine positive
Gomori's aldehyde fuchsin	–
Azo blue-Congo red	Red

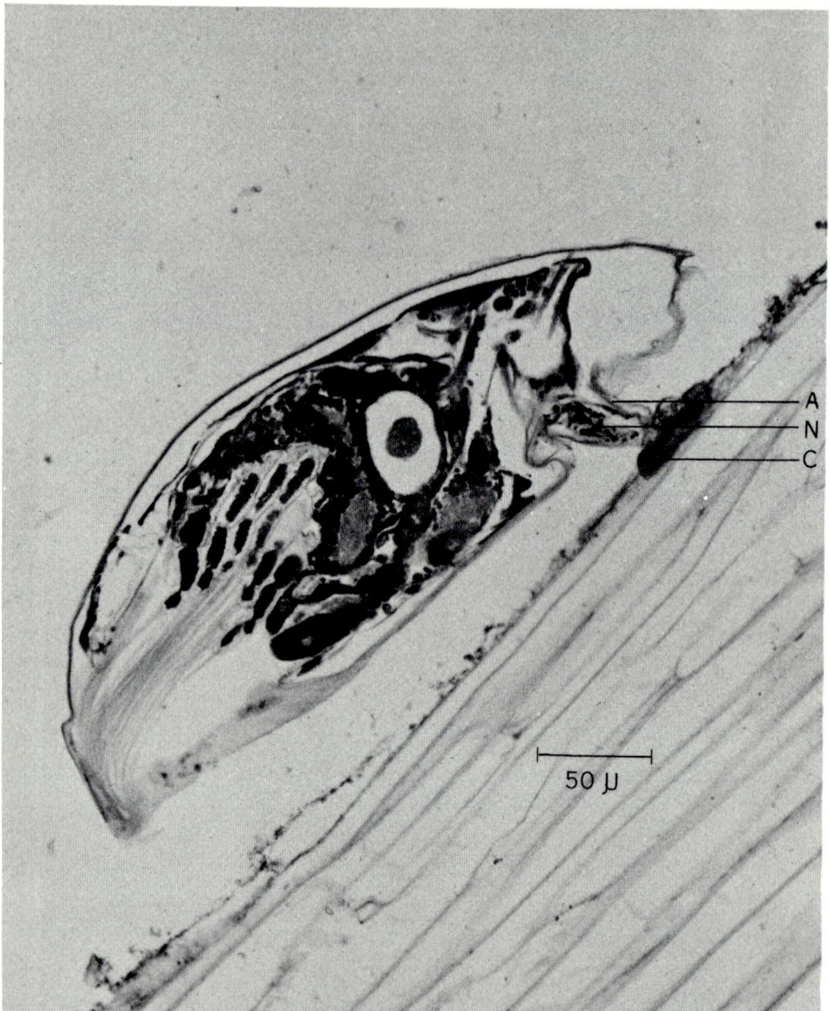

Fig. 1. Microphotograph of a section through an attached cyprid of the barnacle, *B. eburneus*, stained with Mallory's triple stain. A = antenna; N = nerve cells; C = adhesive. The adhesive reacted strongly with the orange G of Mallory's triple stain.

with the acid fuchsin and orange G of Mallory's triple stain. With Wismar's quad-type stain (1966), the red color is apparently due to staining with woodstain scarlet and acid fuchsin, and with the trichrome stain of Green and Wood (1959), the yellow-green color is the result of the adhesive's reaction to fast yellow TN. The acid dye, bromsulfalein, colored the adhesive purple. There was no reaction, however, with the Millon reagents which are specific for tyrosine, and to the Sakaguchi reaction for arginine.

With the specific staining reaction for collagen, ponceau S colored the material red, supposedly indicative of a positive reaction for collagen and reticular fibers.

IV. Discussion

Because there was no reaction with the various reagents for carbohydrate-containing materials and only a trace of positive reaction, if any, with Sudan black B, it can be assumed that the adhesive secreted by the cyprid larvae of the barnacle *Balanus eburneus* is principally a protein or polypeptide. The strong reaction with Biebrich scarlet at a high pH would indicate that the material is a basic protein; however, the Millon and Sakaguchi reactions were negative, thus, ruling out relatively large quantities of the amino acids arginine and tyrosine, often found in basic proteins. The adhesive's reaction with the acid dyes eosin, acid fuchsin, orange G, woodstain scarlet, and the phloxine of Bargmann's chrome hematoxylin reaction also indicates a slightly basic nature.

The reaction with Congo red, usually used to stain cellulose cell walls in plants, where it forms a hydrogen bond with the hydroxyl groups on the cellulose, may possibly be due to the reaction of the dye with a amido group of the protein chain, as suggested by Baker (1958).

The reddish-purple color with Silverman and Glick's bromsulfalein reaction (1966) is difficult to explain unless it be attributed to insufficient exposure of the section to alkaline alcohol following dehydration.

The specific reactions for collagen offer some interesting information on the chemical nature of the adhesive. With the trichrome stain of Green and Wood (1959), the material colored yellow-green, the color for collagen and cytoplasm. The picro-ponceau technique of Gurr (Humason, 1967) stained the adhesive red, also the color for collagen and reticular fibers. In this case cytoplasm should give a yellow reaction. Thus, collagenous material can be differentiated from cytoplasmic protein.

The pale yellow-green color following diazosulfanilic acid is indicative of azo binding sites in the adhesive. This is also confirmed by the staining with other azo dyes such as orange G and Congo red.

Based on the histochemical reactions described above, the barnacle cyprid adhesive appears to be a protein, slightly basic in nature, with a strong affinity for acid dyes. There are no histochemically detectable amounts of phenolic amino acids. The trace of Sudan black B is not enough to permit calling the adhesive a lipoprotein, although this possibility remains open. There are no apparent carbohydrates associated directly with the adhesive.

No ducts through the antennules were discernible with the techniques utilized in this study. The adhesive gave some intense reaction in many cases, but no material which reacted like the adhesive could be seen in any gland or duct. This does not preclude the possibility that the chemical nature of the adhesive, once it has been secreted and hardens, causes it to react differently from the way it would before any molecular cross-linking takes place (if that is how it works), tying up sites which would otherwise be reactive.

Acknowledgments

The authors wish to express their appreciation to Mrs. Darlene Harrington for her skillful preparation of the tissue sections.

References

Arvy, L., and Lacombe, D. (1968). *Compt. Rend.* **267**, 1326–1328.
Arvy, L., Lacombe, D., and Shimony, T. (1968). *Am. Zoologist* **8**, 817.
Baker, J. R. (1958). "Principles of Biological Microtechnique." Methuen, London.
Bernard, F. J., and Lane, C. E. (1962). *J. Morphol.* **110**, No. 1, 19–40.
Bensley, R. R., and Gersh, I. (1933). *Anat. Record* **57**, 217–233.
Bohart, R. M. (1929). *Am. Naturalist* **63**, 353–361.
Clarke, G. L. (1947). *Biol. Bull.* **92**, 73–91.
Crisp, D. J. (1961). *J. Exptl. Biol.* **38**, 429–446.
Crisp, D. J. (1969). Personal communication.
Darwin, C. (1854). "A Monograph of the Cirripedia: The Balanidae." The Ray Society, London.
Doochin, H. D. (1951). *Bull. Marine Sci. Gulf Caribbean* **1**, 15–39.
Green, J. A., and Wood, M. L. (1959). *Stain Technol.* **34**, 313–316.
Groom, T. T. (1894). *Phil. Trans. Roy. Soc. London* **B185**, 119–232.
Herz, L. E. (1933). *Biol. Bull.* **64**, 432–442.
Humason, G. L. (1967). "Animal Tissue Techniques," 2nd ed. Freeman, San Francisco, California.
Kaufmann, R. (1965). *Z. Morphol. Oekol. Tiere* **55**, 161–232.
Lillie, R. D., Palmer, R. W., and Gutierrez, A. (1966). *J. Histochem. Cytochem.* **15**, 421–424.
Lindner, A. (1968). Personal communication.
Michaelis, L. (1947). *Cold Spring Harbor Symp. Quant. Biol.* **12**, 131–142.
Movat, H. Z. (1955). *A.M.A. Arch. Pathol.* **60**, 289–295.

Mowry, R., and Winkler, C. H. (1956). *Am. J. Pathol.* **36**, 628-629.
Nott, J. A. (1969). *Marine Biol.* **2**, 248-251.
Pearse, A. G. E. (1960). "Histochemistry, Theoretical and Applied," 2nd ed. Little, Brown, Boston, Massachusetts.
Silverman, L., and Glick, D. (1966). *J. Histochem. Cytochem.* **14**, 425-426.
Spicer, S. S. (1960). *J. Histochem. Cytochem.* **8**, 18-34.
Spicer, S. S. (1963). *Ann. N.Y. Acad. Sci.* **103**, 322-332.
Spicer, S. S. (1965) *J. Histochem. Cytochem.* **13**, 211-234.
Spicer, S. S., and Meyer, D. B. (1960). *Am. J. Clin. Pathol.* **33**, 453-460.
Visscher, J. P. (1928). *Biol. Bull.* **54**, 327-335.
Visscher, J. P. (1930). *Biol. Bull.* **75**, 341-342.
Wismar, B. L. (1966). *Stain Technol.* **41**, 309-313.

CHAPTER 7

HISTOLOGICAL AND HISTOENZYMOLOGICAL CHARACTERISTICS OF THE CEMENT APPARATUS IN BARNACLES

Vincent R. Liguori

I. Introduction

A. Barnacle Cement

The tenacity with which barnacles remain fixed to most surfaces is an indication of the strong adhesive quality of the cement used for attachment. The bond formed between the cement and the substratum is strong and long lasting and resists the severe stress of water movement constantly present in the marine environment. This natural adhesive apparently meets the criteria for good bonding agents (Sharpe *et al.*, 1964). It is a liquid which spreads easily over a solid surface and, therefore, must have a low level of free energy of the surface; when dried it possesses strong cohesive forces which resist stress within certain limits. It tends to remain intact when dried, indicating it does not undergo undue shrinkage; and since the bond retains its strength, the adhesive appears not to react chemically with the solid substratum.

B. The Barnacle Cement Apparatus

The attachment cement is a product of cellular function within specialized tissues of the barnacle. To be effective, the cement must be conducted from its site of synthesis to the area making contact with the substrate. Barnacles of the family Balanidae are affixed to a solid surface at their basis, while among the Lepadidae, attachment occurs at the base of the peduncle. Several morphologically different cellular structures are involved in barnacle cement synthesis and its conduction to the area of attachment. Because these structures are all involved in fixing the barnacle to a substratum, it seems best to consider them a unit—the cement apparatus.

Darwin (1851), who properly noted the function of the antennae in the initial attachment of the cyprid larva, was the first to describe the cement apparatus in barnacles. He traced the two cement ducts from the terminal discs to two glandular bodies, or "incipient ovaria," at the bases of the antennae. In the mature Lepadidae, he followed the course of the ducts along the muscles within the peduncle to the point where they join the cement glands. When he described the cement apparatus in the Balanidae, Darwin (1854) noted that its greater complexity, compared to that of the Lepadidae, is due to the broad basis. He concluded correctly that the numbers of cement glands and cement conducting canals in the Balanidae increase with periods of growth. The large numbers of cementing structures compensate for the increase in size of the basis by providing sufficient cement at its periphery, thereby insuring attachment at all points which make contact with the substrate. Because of the complexity of the cement apparatus in the Balanidae, Darwin found it difficult to trace the glands and canals in these species. For this reason, he could not confirm, in the acorn barnacles, that the cement glands and their canals consist of modified ovarian tubes, a conclusion he reached from his observations of the stalked barnacles.

After Darwin, several investigators, as a result of their studying the Cirripedia, helped clarify a number of aspects of the barnacle cement mechanism. For example, the exact location of the cement glands and their associated canals in the mantle of the pedunculate barnacle, *Conchoderma*, was first reported by Krohn (1859). Koehler (1888, 1889) and Gruvel (1893, 1905a,b) described the location and the gross morphology of the cement glands and the tubular canals in other Cirripedia, and the role of the cement conducting canals was further clarified by Hoeck (1907) who observed structural differences in the cement complex in three genera, *Lepas*, *Conchoderma*, and *Scalpellum*. Other works dealing with Crustacea (Lankester, 1909; Smith *et al.*, 1909) discussed the adhesive nature of the substance excreted through the antennules of the cypris larva for use in attachment. In spite of the diagrammatical representations in these works which locate the cement glands in the area above the basal plate in the Balanidae and among the ovarian tissues of the peduncle in the Lepadidae, there is no histological description of the structures of the cement apparatus. Krüger (1923) was the first to describe the cement glands of *Scalpellum scalpellum* and *Scalpellum stromii* and to relate variations in the appearance of the cells to a glandular secretory cycle. Johnstone and Frost (1927) histologically characterized the cementing structures of *Anelasma squalicola* in detail and recognized the function of the associated canals. They were not able, however, to locate and describe the nucleus of the cement glands and in spite of their conclusion that the glands manufacture adhesive for attachment, their report leaves no doubt that they were uncertain regarding the proper function of the cementing structures. The cement glands of the Pedunculata and Operculata were regarded by Thomas (1944) to be modifications of the tegumental glands. As a result of his studies, a partial histological description of the cement glands and several of the conducting canals was given for several species of barnacles. The first detailed account of cells, tissues, and structures presumed to be responsible for the synthesis, secretion, and conduction of the cement is provided by Lacombe (1966, 1967), who described a system of canals which give rise to the cement secreting glands in barnacles.

This chapter reviews the studies conducted in Lacombe's laboratory and in our laboratory in a collaborative effort with Lacombe and Arvy on the histological and histoenzymological characteristics of the cement apparatus in barnacles.

II. Experimental Procedures and Observations

A. *Studies with Balanus tintinnabulum*

The arrangement of the canal system in the calcareous basis of *Balanus tintinnabulum* was first described in detail by Lacombe (1966). She infiltrated the basis with a dye under vacuum and found that this treatment enhanced the visibility of the canals and facilitated their study under the microscope. The shell portions of freshly collected barnacles were first decalcified either in 1% hydrochloric acid prepared with seawater or in 1% nitric acid for 30 minutes or longer, depending on the size of the animal. Next, the decalcified specimen, placed in a closed vessel attached to a vacuum source, was suspended on a wire loop above a solution of Trypan blue, Sudan black, or Bismarck brown diluted 1:1000 with a very fine oil. A vacuum of 1/2 atm. was applied for 10 minutes. The barnacle was then submerged in the dye solution by pushing down on the handle of the wire loop which extended through the stopper of the vessel. The canals were filled with the dye within 30 minutes to one hour. Using various histological techniques, she traced the canal system to the ovarian follicles where the cement producing cells are found.

The results of this work indicate that the calcareous basis of *Balanus tintinnabulum* contains radial and circular canals. The two radial canals extend laterally in opposite directions from the points of attachment of the vestigial cypris antennules and connect with the circular canals. The circular canals vary in number according to the size of the animal. They form concentric channels which conduct the cement to the periphery of the basis. Large tubular structures, the principal canals, extend from the ovarian follicles and join the canal system of the basis at the points where the radials join the circular canals. The principal canals penetrate between the ovarian follicles where they divide, giving rise to the smaller, secondary canals.

The cement producing cells, according to Lacombe (1966), reach 40 μ in diameter when mature. The nuclei of the gland cells are polymorphic and contain a small amount of chromatin irregularly scattered along the nuclear membrane. Twelve or more nucleoli may be present. The cytoplasm of the gland cells is differentiated into one zone which has an affinity for a basic dye and one which stains with an acid dye. The difference in staining affinities between the two zones appears to be related to the secretory mechanism.

Histochemical analysis of the cement glands of *Balanus tintinnabulum* (Lacombe, 1967) reveals that the cytoplasm is rich in ribonucleic acids. Microchemical techniques applied in this study to tissue sections and to the extracellular cement at the points of attachment tentatively suggest the presence of acid mucopolysaccharides but not indole compounds (Lacombe, 1968).

B. *Comparison of the Cement Apparatus of Lepas anatifera and Balanus tintinnabulum*

When the histological characteristics of the cementing structures of the acorn barnacle, *Balanus tintinnabulum*, were compared with those of the stalked barnacle, *Lepas anatifera*, several differences and similarities were noticed (Lacombe and Liguori, 1969b). Differences in morphological detail of the cement apparatus of the two species, observed in this study and previously reported by early workers in this field, are attributable to the anatomical differences between the two organisms.

Freshly collected barnacles, fixed in one of several suitable solutions, were decalcified by repeated treatment with Susa's fluid and dehydrated in a progressive series starting with 95% and 100% alcohol followed by alcohol and benzene, (3:1), (1:1), (1:3), (1:7), and absolute benzene before paraffin infiltration. The same dehydration schedule was followed regardless of the fixative used except that Bouin fixed tissue entered the dehydration series after 70% alcohol; formalin and Flemming fixed materials were washed for four hours in running tap water and entered the dehydration series after transfers through 40% and 70% alcohol. Tissue first fixed in Gilson's fluid was treated a second time in Susa's fixative before dehydration.

The paraffin embedded tissues were serially sectioned at 7 μ and stained with one of the following: Delafield's hematoxylin with Chromotrope 2R as a counterstain; Ehrlich hematoxylin counterstained with eosin; nuclear fast red with naphthol green as a counterstain; Heidenhain's iron hematoxylin; or nuclear fast red with azan and Weigert hematoxylin with Congo red and Orange GG used as counterstains.

In both *Lepas anatifera* and *Balanus tintinnabulum*, the cement apparatus is composed of unicellular glands and a series of communicating, tubular canals.

The cement glands of *Lepas anatifera*, found in the peduncle near but distinctly apart from the ovaries, measure 30–40 μ in diameter in an animal 8–10 cm in length. The glands have numerous nucleoli which are surrounded by fine chromatin granules. Each gland cell has a small zone of ergastoplasm near the cell membrane and vacuoles varying in size are found throughout the cytoplasm (Fig. 1). In contrast, the cement gland cells of *Balanus tintinnabulum*, as mentioned above (Lacombe, 1966; Lacombe and Liguori, 1969b), are characterized histologically by a marked polarity in their staining reactions. They may appear singly but more often are found in groups of twenty or more cells that possess a filamentous cytoplasm. In the smaller, young gland cells, the nucleus is usually round but as the cells grow to maturity the nuclei show increasing degrees of polymorphism. Chromatin is concentrated in the small gland cells but is dispersed in the large cells.

In *Lepas anatifera* the canal system is composed of the principal canals, secondary canals, collecting canals, and intracellular canals.

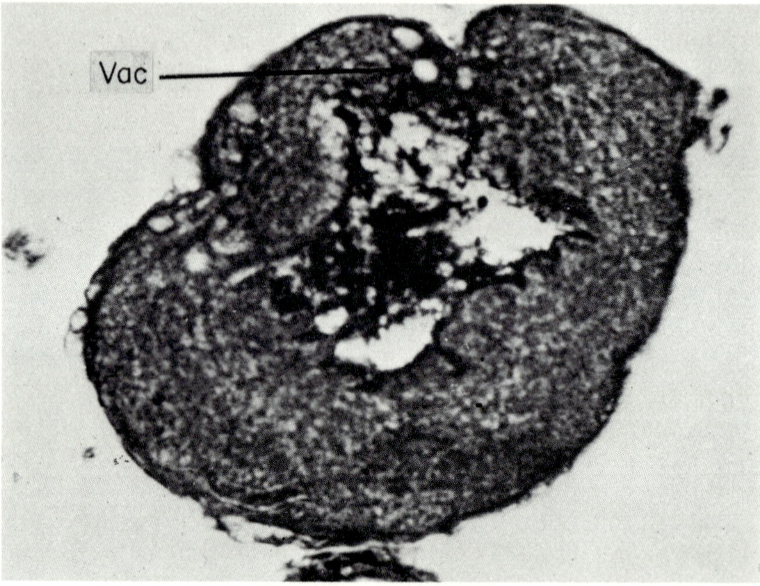

Fig. 1. Individual cement gland of *Lepas anatifera* showing the presence of vacuoles (Vac) in the cytoplasm. ×1000.

The principal canals are two simple ducts that extend from the area of the ovaries to the base by following a course parallel to the walls of the peduncle. The thick walls of the principal canals are lined by a chitinous-like cuticle readily visible under polarized light. Branching of the principal canals occurs among the ovary tissues resulting in the formation of the smaller, secondary canals. The cement gland cells in both species arise from squamous epithelia on the walls of the secondary canals. As an epithethelial cell grows, its nucleus enlarges. As cell growth continues, the walls of the secondary canal penetrate the cytoplasm to form the collecting canal which maintains continuity with the secondary canal. In *Lepas anatifera* the collecting canals ramify to form the intracellular canals, which are lacking in *Balanus tintinnabulum.* On the other hand, *Balanus tintinnabulum* has many principal canals that conduct the cement received from the collecting and secondary canals, to the radial and circular canals in the basal plate.

C. Histoenzymological Studies

1. Alkaline Phosphatase and Succinic Dehydrogenase

Histological studies of the unicellular cement glands of the barnacle

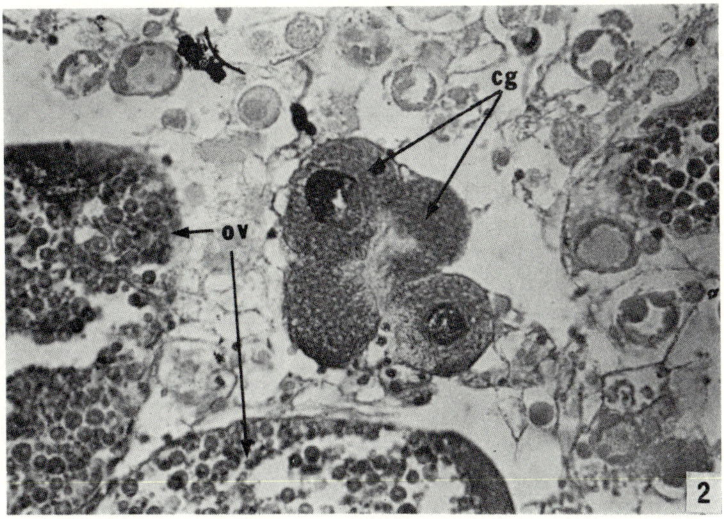

Fig. 2. A cluster of young cement glands (cg) of *Balanus eburneus* among the ovarian follicles (ov). Alcian blue stain (0.1% in 3% acetic acid). ×400.

are often confusing because the gland cells stain with an affinity similar to those of the surrounding ovarian tissues (Figs. 2 and 3). Interest in the enzyme systems of the cementing structures led Arvy and her co-workers (1968) to investigate alkaline phosphatase activity in a pre-

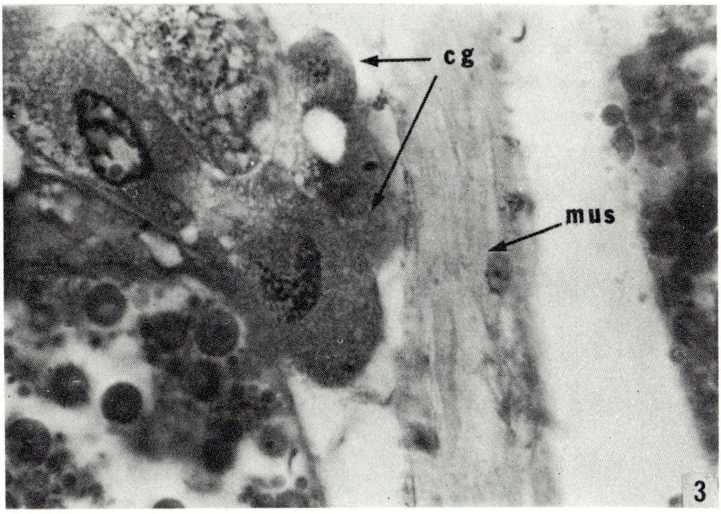

Fig. 3. Cement gland cells (cg) of *Balanus eburneus* near muscle fibers (mus) among the ovarian tissues. Alcian blue stain (0.1% in 3% acetic acid). ×1000.

liminary study of the cement apparatus of *Balanus nubilis*, *Balanus balanoides*, and *Balanus eburneus*. Using the Takamatsu-Gomori technique (Arvy, 1958), they found that the clarity of the cement apparatus was enhanced for microscopic examination by the reaction of the substrate with the enzyme found in the tissues (Figs. 4-7). In this system, the positive alkaline glycerophosphatase activity is selective for

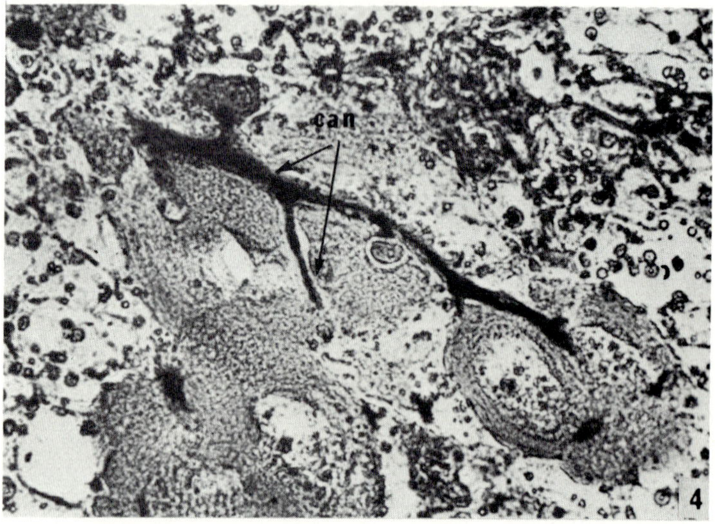

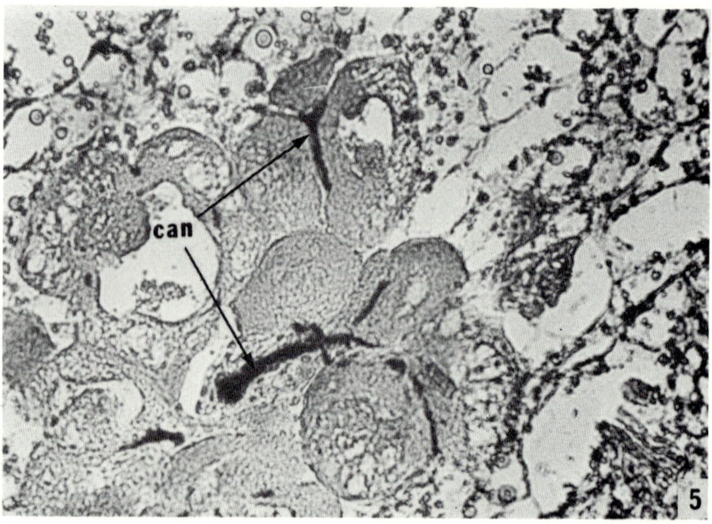

Figs. 4 and 5. Alkaline phosphatase activity localized in the cement conducting canals (can) of *Balanus nubilis*. ×400.

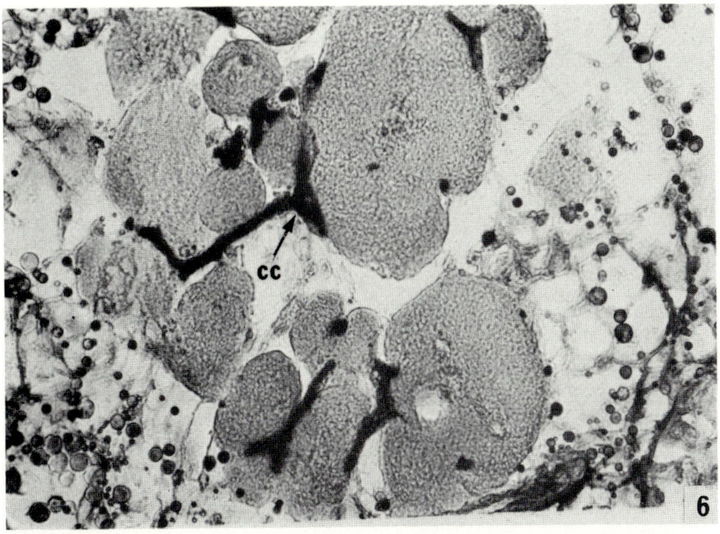

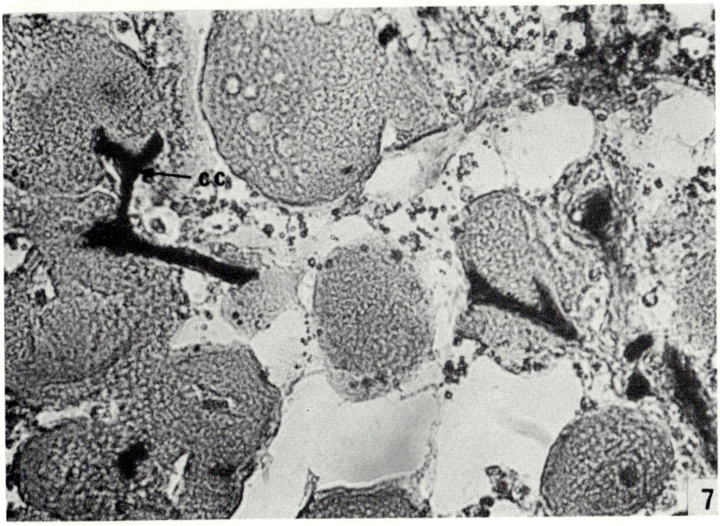

Figs. 6 and 7. Alkaline phosphatase activity localized at the pole of cement accumulation (cc) and in the canals of *Balanus nubilis*. ×400.

the cement apparatus in all phases of its development, active secretion and regression. The results show that the preductal poles of the gland cells and the walls of the canals and their contents are rich in the enzyme (Figs. 8 and 9). When the tissues are incubated in an α-naphthol phosphoric acid bath (with 0.1 *M* sodium veronal buffer at

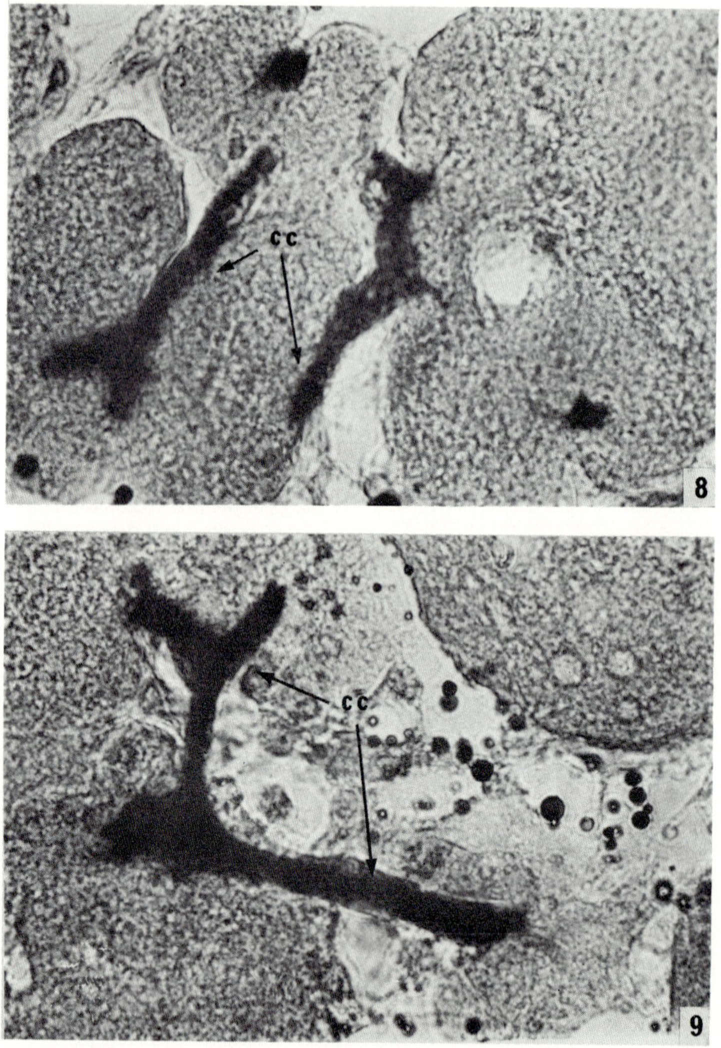

Figs. 8 and 9. Positive alkaline phosphatase activity in the cement conducting canals (cc) of *Balanus nubilis,* shown at a higher magnification. ×1000.

pH 9.2 coupled with fast blue RR), only the excretory canal system of the glands are visible.

In a later study, Arvy and Lacombe (1968) examined the alkaline phosphatase and succinic dehydrogenase activities in the cement apparatus of the same three species. The animals, removed from their shells, were fixed for several hours at 4°C in either 10% neutral formalin, 95% alcohol, or acetone. After paraffin inclusion, the tissues were sectioned at 10–12 μ and incubated in suitable substrates to reveal alkaline phosphatase activity. Succinic dehydrogenase activity was examined in ovary and mantle tissues incubated in a bath containing sodium succinate and nitro BT in a 0.1 *M* Sørensen buffer at pH 7.6. After incubation for one hour the tissues were fixed in formalin, embedded in paraffin, and sectioned at 8–10 μ.

The results of this study show that the cytoplasm of the young gland cells is rich in alkaline glycerophosphatase, but that in the mature cells enzyme activity is more detectable in the area of the collecting canals. In addition, alkaline phosphatase activity is found to be high in all of the canals closely associated with the gland cells. Succinic dehydrogenase activity was also found to be present in high concentrations in the young gland cells but was not histochemically detectable in the canal system.

2. Cytochrome Oxidase

Cytochrome oxidase activity was not histochemically detectable in the cement apparatus of *Balanus nubilis*, *Balanus balanoides*, or *Balanus eburneus* in a preliminary study of the distribution of this enzyme in barnacle tissues (Arvy and Liguori, 1968). Enormous differences were found, however, in the activity of this enzyme in the muscles of these animals.

Freshly collected live barnacles were removed from their shells and rapidly dissected and the tissues incubated in a substrate according to the method of Burstone (1960) with *N*-phenyl-*p*-phenylenediamine in 0.2 *M* Tris buffer at pH 7.4 using sodium azide as a control.

Cytochrome oxidase activity was not detectable in the large, white, striated depressors, nor in the rostral, tergal, and lateral muscles; but the muscles of the thoracic appendages, mandibles, and maxilla are rich in the enzyme. The motor muscles of the terminal cirri rapidly stain a dark blue when incubated in Burstone's bath, an indication that these tissues are particularly rich in this enzyme.

III. Discussion

The more primitive stalked barnacles represented by *Lepas anatifera* have the cement apparatus in the peduncle close to the ovarian follicles. Cement formed in the gland cells is conducted through a series of tubular canals to the base of the peduncle where it apparently hardens after contact with the substrate. In the *Balanidae*, especially those members of the acorn barnacles with a calcareous basal plate like that of *Balanus tintinnabulum,* cement conducted through the canal system passes into the circular canals at the base before it makes contact with the substrate.

Among closely related species of the Lepadidae, variations are evident in the morphology and distribution of the cement glands and their canals. Koehler (1889) reported that the cement producing cells of *Lepas* are found in the peduncle 2–3 mm beneath the capitulum. The unicellular glands of *Scalpellum* extend in the peduncle for a greater distance, but in *Pollicepes* they are distributed along its entire length to the base. In *Conchoderma* the cement secreting cells were found by Krohn (1859) to be situated in the mantle, but in the parasitic barnacle *Anelasma squalicola* neither the glands nor their canals were found in the mantle, thorax, or parts of the body occupied by the ovarian follicles. Instead, groups of closely packed glands were located in the region of the waist on either side of the two primary canals (Johnstone and Frost, 1927).

The distribution of the cement glands may also vary among closely related species in the Balanidae (Lacombe and Liguori, 1969a). In *Balanus tintinnabulum,* for example, the cement glands are distributed among the interfollicular spaces of the ovaries and the connective tissue surrounding the ovarian follicles. But in *Balanus nubilis,* a related species, the glands and their canals are not so closely arranged in the interfollicular spaces.

The cement glands of barnacles are well known, and their gross morphology has been described by several earlier investigators (Darwin, 1851; Gruvel, 1905a; Krüger, 1923; Johnstone and Frost, 1927). There are, however, basic cytological differences in the cement gland cells of *Lepas anatifera* and *Balanus tintinnabulum* which may indicate differences in their secretory mechanisms. Actively secreting gland cells of *Lepas anatifera* have many nucleoli, a small zone of ergastoplasm near the cell membrane, and large numbers of vacuoles, varying in size, distributed throughout the cytoplasm (Lacombe and Liguori, 1969b). These are characteristics of cells having a high level of metabolic activity. It is interesting to note that Thomas (1944), in his brief account of barnacle cement glands, described actively secreting cells in *Lepas anatifera* as having a finely granular cytoplasm without vacuoles and a nucleus with one nucleolus. These characteristics adequately describe the immature gland cell of *Lepas anatifera,* which presumably has not yet attained the ability to secrete actively. Thomas (1944) noticed variations in the appearance of the cells which he related to changes that occur as a result of a secretory cycle. Since he provides no description of the cell variations, it seems plausible to suspect that these were, in fact, actively secreting mature cells which he observed. In the case of *Lepas anatifera,* the cement glands correspond to the apocrine type cell in which secretory function is accompanied by a loss of some cytoplasm. As a result of secretory function, variations in the appearance of the cells may be evident. The histological evidence indicates that in *Lepas anatifera* the cellular secretion is transported to the intracellular canals in vacuoles formed in the cytoplasm. The syncytial nature of the walls of the intracellular canals, collecting and secondary canals, may enhance transport of the soluble secretion across the cell membranes.

Unlike that of *Lepas anatifera,* the cytoplasm of the mature gland cells of *Balanus tintinnabulum* is filamentous rather than vacuolated. The gland cells have polymorphic nuclei and may contain twelve or more nucleoli. However, these glands correspond to the merocrine type cell in which secretory function results in no loss of cytoplasm. When tissue sections are stained with nuclear fast red using naphthol green as a counterstain, the more dense zone of the cell, near the collecting canal, stains green while the less dense zone at the opposite pole stains red. The difference in the staining reaction at opposite poles indicates a difference in pH. The histological evidence suggests that the cellular secretion formed at the far pole is transported in the cytoplasm toward the collecting canal where it accumulates. The change in pH between the poles of the cell may account for an increase in solubility of the cellular product and its passage across the membrane of the collecting canal.

The principles of localizing enzymes in tissue sections involve (1) finding a suitable substrate which can be converted to a visible, colored compound as a result of enzyme action and (2) finding suitable conditions in which the reaction can take place. These principles are satisfied when standardized histoenzymological techniques are applied to the study of barnacle tissues.

Interpretation of histological data of barnacle cement glands is often difficult because the surrounding ovarian tissues stain with an affinity similar to that of the cement glands. The deposition of colored compounds resulting from enzyme action within the tissues clearly defines the structural characteristics of the tissue components studied. The positive alkaline phosphatase activity in the tissues of the cementing structures confirms the histological findings regarding the disposition of the canal system. The positive enzyme activity also is an indication of the high metabolism of the cementing structures in the formation and transportation of the cement. This seems apparent since alkaline phosphatase is thought to function in the maintenance of intracellular inorganic phosphate, in dephosphorylation necessary for absorption and transport, and as a metabolic control by dephosphorylation.

The positive succinic dehydrogenase activity histochemically detectable in the young gland cells also suggests a high metabolic level for these tissues, probably the result of oxidation-reduction reactions providing energy for cell growth.

One of the final steps in the respiratory chain mechanism in cells is the process in which cytochrome c is reoxidized with the ultimate formation of water from the combination of hydrogen with oxygen. This step is catalyzed by cytochrome oxidase. This enzyme, histochemically detectable in several barnacle muscles, was found to correlate well with muscle activity. The more active motor muscles of the terminal cirri have a greater cytochrome oxidase complement than any of the other barnacle muscles studied. As a result of our investigations of barnacle tissues, it seems most probable that specialized muscles play a major role in the conduction of cement through the canals to the points of attachment (Lacombe and Liguori, 1969a) (Fig. 10).

IV. Conclusions

The examination of the histological and histoenzymological charac-

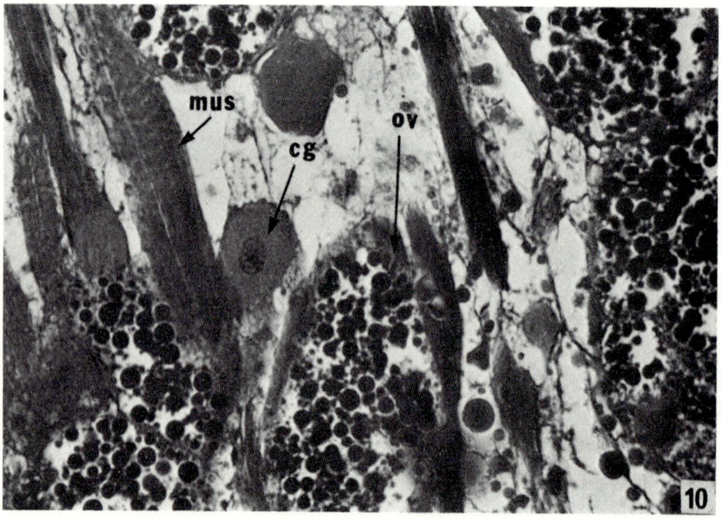

Fig. 10. Young cement gland (cg) of *Balanus eburneus* among the ovarian follicles (ov) and the muscle fibers (mus). Heidenhain's iron hematoxylin. ×400.

teristics of barnacle tissues has helped to clarify the cement forming and conducting mechanisms in these animals. At present, studies being conducted on the fine structure of the cement apparatus (Lacombe, 1968) may further clarify the mechanisms of secretion and conduction. It is hoped that these studies and histoenzymological investigations of the cement apparatus currently in progress in this laboratory may lead to the chemical identification of the adhesive material secreted by barnacles.

Acknowledgments

Some of these studies have been supported by grants from the Office of Naval Research, Department of the Navy (Contract N00014-68-C-0334); the US Naval Applied Sciences Laboratory, Brooklyn, New York (Contract N00014-69-C-0260); and the Rockefeller Foundation (RF 64078) to the Osborn Laboratories of Marine Sciences.

The author is grateful to Dr. Ross F. Nigrelli, Director of the Osborn Laboratories of Marine Sciences and the New York Aquarium; Dr. George D. Ruggieri; and Mrs. Tamar Shimony for their critical appraisal of the manuscript.

References

Arvy, L. (1958). *Extrait Biol. Med.* **46/47**, 4–8.
Arvy, L., and Lacombe, D. (1968). *Compt. Rend.* **267**, 1326–1328.
Arvy, L., and Liguori, V. R. (1968). *Am. Zoologist* **8**, 383.
Arvy, L., Lacombe, D., and Shimony, T. (1968). *Am. Zoologist* **8**, 384.
Burstone, M. S. (1960). *J. Histochem. Cytochem.* **8**, 63–70.
Darwin, C. (1851). "The Lepadidae or Pedunculated Cirripedes." The Ray Society, London.
Darwin, C. (1854). "The Balanidae, The Verrucidae, etc." The Ray Society, London.
Gruvel, A. (1893). *Arch. Zool. Exptl. Gen.* [3] **1**, 401–610.
Gruvel, A. (1905a). *Zool. Jahrb.* Suppl. 6, 307–352.
Gruvel, A. (1905b). "Monographie des Cirrhipèdes ou Thécostracés." Masson, Paris.
Hoeck, P. P. C. (1907). *Siboga Exped.* **31**, 1–127.
Johnstone, J., and Frost, W. E. (1927). *Lancashire Sea Fisheries Lab. Rept.* No. 35, 29–31.
Koehler, R. (1888). *Rev. Biol. Nord France* **2**, 41–55.
Koehler, R. (1889). *Arch. Biol. (Liege)* **9**, 311–402.
Krohn, A. (1859). *Arch. Naturgesch.* **25**, 355–364.
Krüger, P. (1923). *Arch. Mikroskop. Anat. Entwicklungsmech.* **97**, 839–872.
Lacombe, D. (1966). *Publ. Inst. Pesquisas Marinha, Nota Tecnica* No. 32, 1–39.
Lacombe, D. (1967). *Publ. Inst. Pesquisas Marinha* No. 011, 1–29.
Lacombe, D. (1968). *Publ. Inst. Pesquisas Marinha* No. 017, 1–22.
Lacombe, D., and Liguori, V. R. (1969a). Unpublished results.
Lacombe, D., and Liguori, V. R. (1969b). *Biol. Bull.* **137**, 170–180.

Lankester, E. R. (1909). "A Treatise on Zoology," Part VIII. Adam & Black, London.
Sharpe, L. H., Schonhorn, H., and Lynch, C. J. (1964). *Intern. Sci. Technol.* **28**, 26–37.
Smith, G., Woods, H., Shipley, A. E., Warburton, C., and Thompson, D. W. (1909). "Crustacea and Arachnids. The Cambridge Natural History." Macmillan, New York.
Thomas, H. J. (1944). Tegumental glands in the Cirripedia Thoracica. *Quart. J. Microscop. Sci.* **84**, 257-282.

CHAPTER 8

COMPOSITION OF MUSSEL AND BARNACLE DEPOSITS AT THE ATTACHMENT INTERFACE

M. Cook

I. Collection of Bioadhesive Material

A. *Mussel Bioadhesive*

The bioadhesive secretion from the common sea mussel forms the attachment disc which attaches the distal end of the byssus thread to the adherend (Fig. 1). Attachment discs were collected on a plate of polytetrafluorethylene (PTFE) 50 cm square and placed in a tank of continuously flowing sea water containing 2 inch–2½ inch specimens of the common sea mussel, *Mytilus edulis*. These were left for inter-

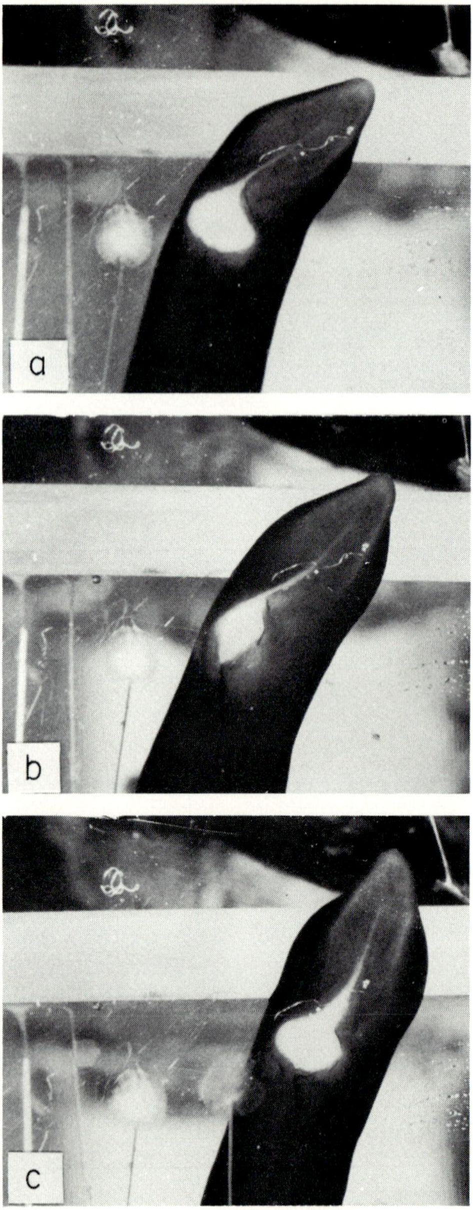

Fig. 1. Series of photographs showing the mode of secretion of the bioadhesive attachment disc by *Mytilus edulis*. (a) The rim of the distal pit is dilated during the secretion of the attachment disc, the light area being formed by the internal surface of the pit. (b) As the foot moves away, the rim of the distal pit contracts. (c) The attachment disc secreted in (a) is now apparent while another disc is being secreted to the right of the photograph.

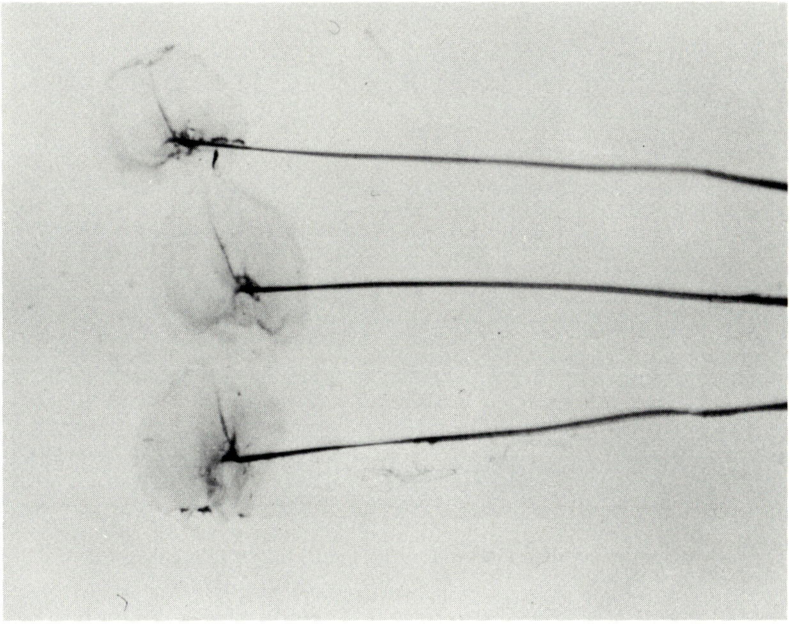

Fig. 2. Distal end of byssus threads of *Mytilus edulis* bonded to PTFE adherend by the attachment discs.

vals of 1 week, after which the animals were detached by manual tensile force which caused the byssus threads to break, leaving the distal end of the threads still attached to the PTFE adherend by the attachment discs (Fig. 2). The plate was then well washed with distilled water and allowed to dry at room temperature. The attachment discs were gently brushed off the dry PTFE plate (Fig. 3) and the outer portions removed for analysis, thus avoiding contamination of the sample with the end of the byssus threads. Similar procedures were followed for collection of attachment discs of the horse mussel, *Modiolus modiolus*.

B. Barnacle Bioadhesive

The species of barnacle used in these investigations was *Balanus crenatus*, which is a member of the family Balanidae and which has a calcified basis. Specimens attached to the periostracum of mussel shells were obtained from the Menai Straits in North Wales and Millport, on the Isle of Cumbrae, in Scotland. The barnacle was detached from the mussel shell by a lateral and slightly upward pressure to the edge of the basis (Figs. 4 and 5). By this method it was possible

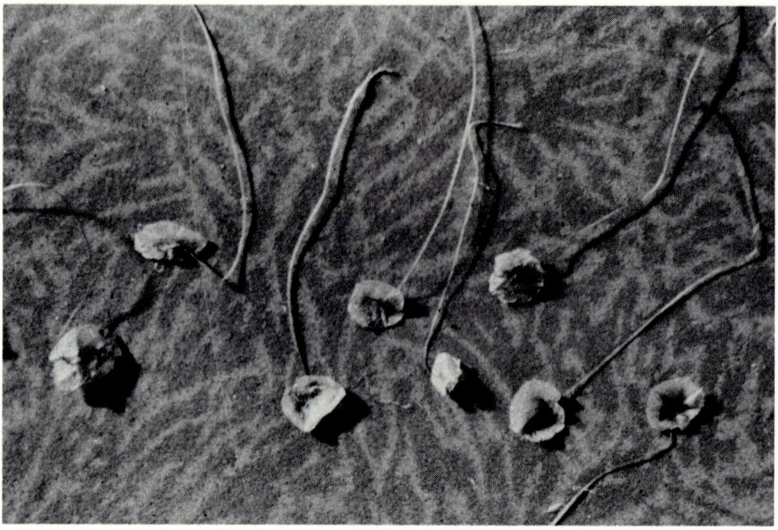

Fig. 3. Detached attachment discs of *Mytilus edulis* from PTFE adherend.

to detach about 50–70% of the barnacles without damaging the basis. The dark coloration of the periostracum of the mussel shell made it easy to detect and to eliminate any contamination from this adherend. The bioadhesive was collected from the basis by peeling with a scalpel. Solitary specimens, when detached, have only a very thin layer of bioadhesive (Fig. 4), whereas in crowded specimens this layer is considerably thicker (Fig. 5), especially when the basis of one specimen has grown up the parietal plate of another.

II. Analytical Methods

A. *Microcombustion Analysis*

After extraction of lipid content with chloroform-methanol (1:1) the sample was subjected to microcombustion analysis using the Pregl method for carbon and hydrogen and the Dumas method for nitrogen determinations. The microcombustion analysis was carried out by the Alfred Bernhardt Mikroanalytisches Laboratorium, West Germany.

B. *Amino Acid Analysis*

Material, dried over P_2O_5, was hydrolyzed with 6 *N* HCl under vacuum at 105°C for 6 hours. The amino acids present in the hydrolyzate

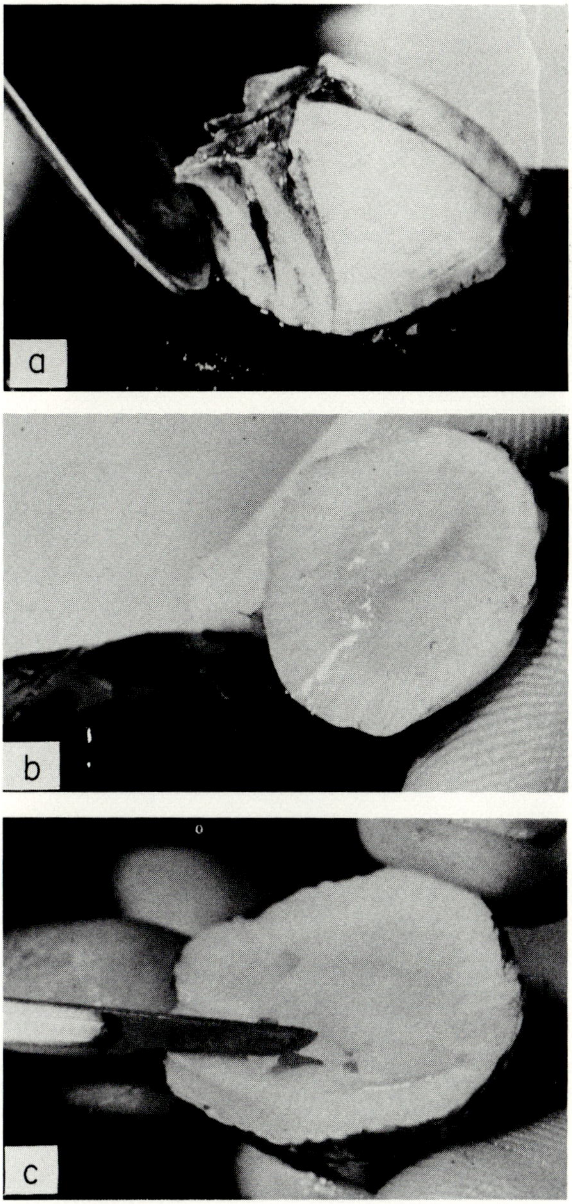

Fig. 4. Detachment and collection of bioadhesive from solitary specimen of *Balanus crenatus* attached to mussel shell. (a) Application of lateral stress to the edge of the basis; (b) detached specimen with basis intact; and (c) peeling off the bioadhesive from the basis with a scalpel.

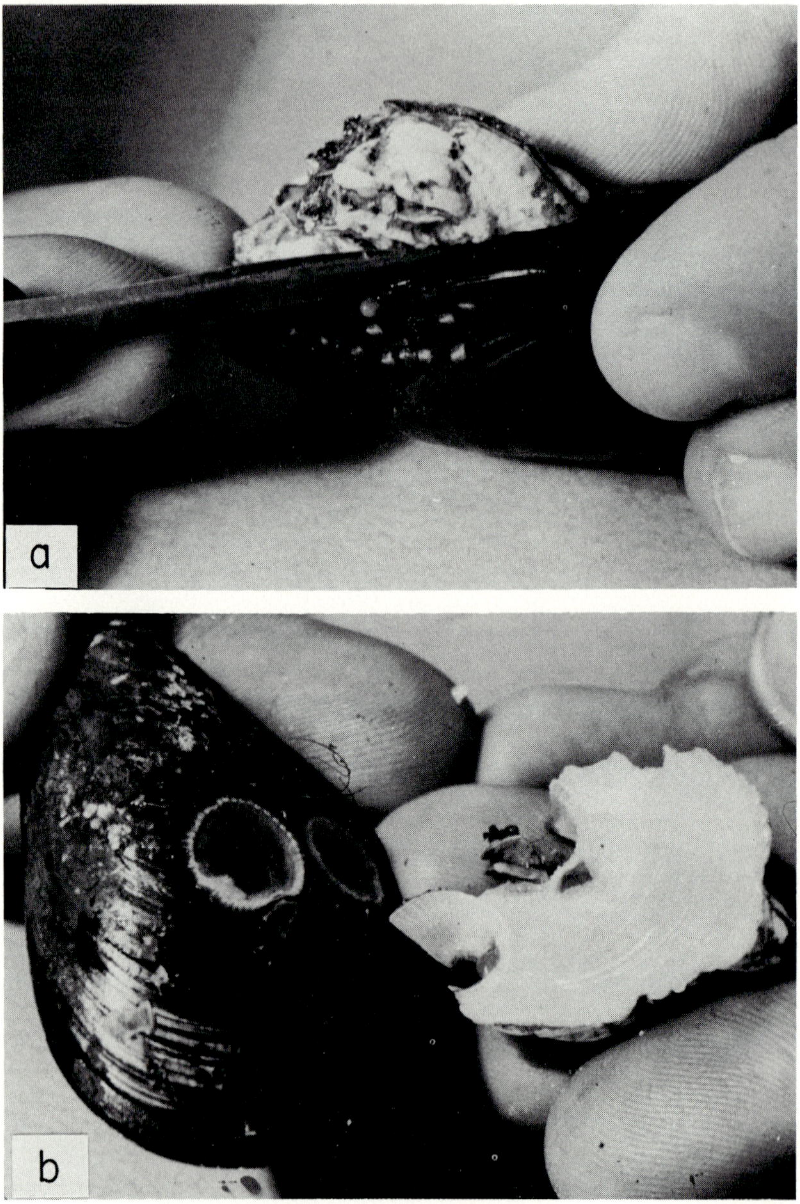

Fig. 5. Detachment and collection of bioadhesive from a group of specimens of *Balanus crenatus*. (a) Application of lateral stress to the edge of the basis; (b) detached specimens; and (c) layer of bioadhesive much thicker compared with Fig. 4 (c).

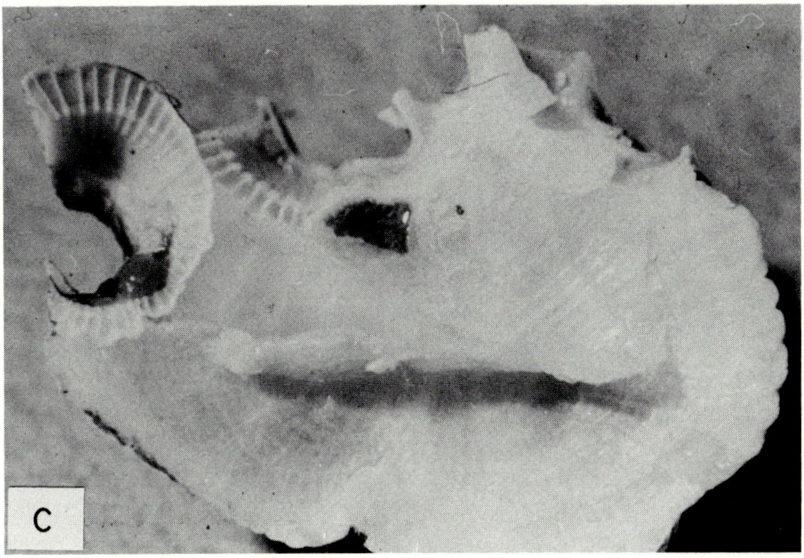

Fig. 5 (c).

were determined on a Technicon Amino Acid Autoanalyser (Stein and Moore, 1954; Thomson and Miles, 1964). Hydrolysis losses have not been assessed but are thought likely to be small under these conditions. Glucosamine, which is likely to be lost under such conditions, did, in fact, show up on the barnacle tracings. However, the figures in Table II for cystine and tryptophan have been left blank. The amino acid analyses were carried out by A. E. Thompson in the Department of Organic Chemistry at the University of Newcastle upon Tyne.

C. *Carbohydrate Analysis*

Qualitative analyses were carried out by descending paper chromatography and developed in ethyl acetate-acetic acid-water-pyridine (5:1:3:5). Quantitative determinations were performed by assaying with Nelson's alkaline copper reagent (Nelson, 1964). Samples for quantitative analysis of neutral sugars were hydrolyzed with 0.5 *N* H_2SO_4 at 105°C for 3 hours and for reducing amino sugars with 2 *N* HCl at 105°C for 22 hours, the HCl and H_2SO_4 being removed before analysis. These analyses were carried out by B. J. Smith in the Department of Biochemistry at the University of Newcastle upon Tyne.

D. Lipid Analysis

Total lipid content was determined, on samples dried over P_2O_5, by direct weighing of chloroform-methanol (1:1) extract.

An initial characterization of the lipid content of the barnacle bioadhesive has been carried out by I. Ross in the Department of Agricultural Biochemistry at the University of Newcastle upon Tyne.

The technique used was thin layer chromatography (TLC) on plates coated with 0.5 mm layer silica G. Plates were developed for:

(a) Neutral lipids in diethyl ether-hexane-ethanol-acetic acid (40:50:2:0.2) for 15 cm and diethyl ether-hexane (6:94) for 20 cm. Location was determined by spraying plates with 0.6% potassium dichromate in 30% w/w H_2SO_4 and charring at 180°C for 20 minutes.

(b) Polar lipids in chloroform-methanol-acetic acid-water (65:15:10:4). One plate was sprayed with 0.5% ninhydrin in butanol-acetic acid (100:3) and heated at 120°C for 5 minutes for location of amino groups. The plate was then resprayed with molybdate reagent (Vashovsky and Kostetsky, 1968) for location of phosphorus. A second plate was treated with periodate-Schiff reagent of Shaw (1968) to detect α-glycols.

Gas-liquid chromatography (GLC) was used to determine the fatty acid composition. After methylation of the lipid extract, separation was carried out on a 7 foot × ¼ inch glass column packed with 15% polyethyleneglycol succinate on 85% acid-washed Chromasorb W at 178°C with a nitrogen flow rate of 50 ml/min. Fatty acids were detected with a flame ionization detector. As a complementary technique to direct gas-liquid chromatography the remaining fatty acid methyl esters were separated by thin layer chromatography on plates coated with 0.5 mm layer of silica G prepared with 10% $AgNO_3$ solution. The developing system was hexane-benzene-diethyl ether (49:49:6). Nondestructive location was carried out by UV fluorescence after spraying plates with ammoniacal Rhodamine-6-G. The relevant areas were scraped off and eluted with 3 × 2 ml chloroform-methanol (2:1). After being dried, the eluate was redissolved in 2 ml methanol-water (1:1) and extracted with 3 × 1 ml of hexane. The volume was then reduced to 0.01 ml by drying with Na_2SO_4 (anhyd), and the various fractions were subjected to gas-liquid chromatography to compare them with the original fatty acid methyl ester profile.

E. Electron Microprobe Analysis

Analysis of the attachment disc of *Mytilus edulis* bonded to 24 ct.

gold adherend was carried out by J. Ibbetson at the Department of Geophysics at the University of Newcastle upon Tyne.

III. Results of Chemical Analysis

A. Mytilus edulis

Microcombustion analysis (Table I) indicates that the bioadhesive is mainly protein in nature. Seventeen amino acids were identified in the acid hydrolyzates (Table II). Glycine represented one-fifth of the total amino acids. Slightly more basic than acidic amino acids and relatively high proportions of aromatic amino acids were also present. Similar figures were obtained for *Modiolus modiolus* bioadhesive (Table II).

The lipid content was 8% by dry weight, and determination of phosphorus content in this extract showed only 0.08% to be present. However, electron microprobe analysis showed approximately 0.2% of phosphorus and also identified the presence of sulpher and calcium. No detectable amount of carbohydrate was found.

B. Balanus crenatus

Sixteen amino acids in all were identified in the acid hydrolyzates (Table III). A tenth of the residues are of glycine and the proportions of hydroxy, acidic, and basic amino acids are similar. The nonpolar amino acids account for approximately 40% of the residues, and again there are relatively high figures for the aromatic amino acid residues.

Glucose and glucosamine were identified by paper chromatography and quantitative determinations gave 2.4% of reducing amino sugars and less than 0.2% glucose.

The lipid content was 27.4% by weight. The major neutral components appear to be hydrocarbons, triglycerides, and free fatty acids with some free and esterified sterols as minor components. There was no evidence of polar lipids being present, but this finding is inconclusive because of the small amount available for analysis and the length of time between collection and analysis. Lipids containing amino groups and phosphorus were identified but there were no indications of any α-glycol-containing compounds.

Eighteen peaks were present on the GLC tracing, of which four were major components amounting to 93% of the fatty acid content. One of these is stearic acid but the identification of the other three is by no means certain; they have been tentatively identified as $C_{20,}$ $C_{23:1}$

TABLE I
MICROCOMBUSTION ANALYSIS OF COMMON SEA MUSSEL *Mytilus edulis* BIOADHESIVE

Constituents	Percentage
Carbon	43.35
Hydrogen	6.73
Nitrogen	15.91
Residues	5.3

TABLE II
AMINO ACID COMPOSITION OF MUSSEL BIOADHESIVE EXPRESSED AS RESIDUES OF AMINO ACID PER 100 RESIDUES

Amino acids	Animals and the location of their collection			
	Mytilus edulis			*Modiolus modiolus*
	Newbiggin	Blyth	Menai Bridge	Port Erin
Hydroxy proline	3.5	3.8	2.4	2.8
Aspartic	9.4	8.4	9.3	10.5
Threonine } Serine }	11.1	11.0	10.6	11.2
Glutamic acid	5.2	3.8	5.4	5.6
Proline	5.6	3.6	5.4	4.4
Glycine	20.1	21.6	18.7	22.4
Alanine	8.7	9.9	8.8	8.4
Valine	4.2	4.2	4.7	6.2
Methionine	0.3	trace	0.6	0.7
Isoleucine	2.6	2.4	1.7	1.7
Leucine	5.2	4.9	5.2	3.7
Tyrosine	5.6	4.2	4.9	3.8
Phenylalanine	3.8	3.7	3.5	2.6
Lysine	5.8	4.1	6.0	6.6
Histidine	5.6	4.9	5.4	3.2
Arginine	8.0	6.7	7.4	7.0
Cystine	—	—	—	—
Tryptophan	—	—	—	—

and *trans* $C_{26:1}$. The latter is the major fatty acid accounting for 56% of the total; there is, however, the possibility of an additional group on the molecule as the retention time on the TLC and GLC seem slightly displaced from what one might have predicted for *trans* $C_{26:1}$.

TABLE III
AMINO ACID COMPOSITION OF BARNACLE BIOADHESIVE AND LOBSTER FIBRINOGEN EXPRESSED AS RESIDUES OF AMINO ACID PER 100 RESIDUES

Amino acids	Species and location of collection: *Balanus crenatus*		Lobster fibrinogen[a]
	Menai Bridge	Mill Port	
Hydroxy proline	0	0	—
Aspartic acid	10.1	8.9	10.1
Threonine } Serine }	15.1	5.0	7.9
		5.1	7.2
Glutamic acid	7.1	8.8	12.8
Proline	6.6	6.0	5.5
Glycine	10.1	8.9	5.2
Alanine	6.6	7.2	5.0
Valine	5.0	7.5	7.1
Methionine	trace	0.7	1.3
Isoleucine	3.3	5.2	5.6
Leucine	8.7	9.0	9.4
Tyrosine	5.9	5.8	3.5
Phenylalanine	4.8	4.9	5.5
Lysine	6.1	6.9	4.9
Histidine	2.7	2.7	3.5
Arginine	7.9	6.9	4.0
Cystine/2	—	—	1.2
Tryptophan	—	—	0.02

[a]Values for lobster fibrinogen taken from Table IV, Stewart *et al.*, 1966.

IV. Discussion

The bioadhesives of both classes are mainly protein in nature with a minor lipid component, the latter being considerably greater in the barnacle bioadhesive. The amino acid composition (Tables II and III) of the two classes shows certain differences: (1) larger amount of glycine residues in the mussel bioadhesive; (2) larger amount of nonpolar amino acids in the barnacle bioadhesive; and (3) absence of hydroxy proline in the barnacle bioadhesive.

An interesting point from the amino acid figures is the close similarity between those for the barnacle bioadhesive and those for crustacean fibrinogen (Table III).

The small amount of glucose and glucosamine in the barnacle bioadhesive is thought likely to be due mainly to contamination of the sample from the outer chitinous cuticle of the basis (Costlow, 1956) and the organic (bacterial) plaque which is present on all objects placed in the oceanic environment.

Acknowledgments

I wish to express my gratitude to Mr. A. E. Thompson, Dr. B. J. Smith, Dr. J. Ibbetson, and Dr. I. Ross for carrying out the chemical analyses and to the Director of the Dove Marine Laboratory, Cullercoats, for laboratory and seawater tank facilities. Also my appreciation to Mr. H. T. Oliver for producing the illustrations and to the Science Research Council for a grant in support of this work.

References

Costlow, J. (1956). *J. Morphol.* **99**, 359.
Nelson, N. J. (1944). *J. Biol. Chem.* **153**, 375.
Shaw, N. (1968). *Biochim. Biophys. Acta* **164**, 435.
Stein, W. H., and Moore, S. (1954). *J. Biol. Chem.* **211**, 915.
Stewart, J. E., Dingle, J. R., and Odense, P. H. (1966). *Can. J. Biochem.* **44**, 1447.
Thomson, A. R., and Miles, B. J. (1964). *Nature* **203**, 438.
Vashovsky, V. E. and Kostetsky, E. Y. (1968). *J. Lipid Res.* **9**, 396.

PART III

ADHESIVES FOR SOFT TISSUE

CHAPTER 9

CHEMICAL MECHANISMS SUITABLE FOR BONDING TISSUES

Gerald A. Grode, Richard D. Falb, C. William Cooper,
and
Jackson Lynn

During the past 15 years more investigators have become interested in performing research on substances which might be used as adhesives for human tissues. This chapter discusses some of the possible mechanisms by which strong adhesive bonds might be formed between a potential adhesive material and human tissue. In particular, primary chemical bonding between tissue and adhesive is discussed.

In general, three methods exist by which adhesive bonds to tissue might be effected. These are: (1) physical or mechanical bonds; (2) secondary chemical bonds; and (3) primary, ionic, or covalent chemical bonds.

I. Physical or Mechanical Bonds

Physical or mechanical bonds are those formed by the inclusion or "locking in" of a material within the cracks or crevices of a substrate. Since this could, of course, occur on a molecular level, tissue seems to present ample opportunities for this type of bonding. Materials which could function well by this kind of mechanism would be highly fluid products, such as cyanoacrylate adhesives, able to penetrate into the interstices of tissue and become immobilized by polymerization. Bond strengths, of course, would depend on the morphology of the tissue site and on the inherent mechanical properties of the specific adhesive material.

II. Secondary Chemical Bonds

Secondary chemical bonds are those attributed to forces such as hydrogen bonding or van der Waals' attractions. These are secondary valence forces, some of which depend on the chemical nature of the two adhering molecules. A universal component of the van der Waals' forces is the London dispersion force which acts between molecules regardless of their chemical characteristics. These forces are of sufficient magnitude to contribute significantly to the strength of some adhesive joints.

III. Primary Chemical Bonds

Primary bonds are obtained by the chemical reaction of an adhesive material with the substrate. The limiting strength of such bonds might approach the high inherent strengths of interatomic bonds, such as carbon-carbon and carbon-nitrogen. This discussion is directed only to aspects of materials capable of forming bonds of this nature.

Recently, a great deal of work has been carried out on the insolubilization of enzymes by chemical reactions of these materials with inert substrates. A wide variety of these reactions have been reviewed by Katchalski and Silman (1966). Since enzymes are, of course, proteins, the binding agents used in these studies would be of interest as possible surgical adhesives. In much of this work the binding reactions were performed under "mild" conditions to prevent inactivation of the specific enzymes.

A number of functional-chemical groups suitable for covalent bonding are present in proteins. These functional groups include α- and ϵ-amino groups; α-, β-, and γ-carboxyl groups; the sulfhydryl and hydroxyl groups of cysteine and serine; the imidazole group of histidine; and the phenol ring of tyrosine. Amino groups readily react with acylating and alkylating agents, with aldehydes, isocyanates, and diazonium salts. Hydroxyl groups both of serine and tyrosine may also react with acylating agents. Sulfhydryl groups react readily with alkylating agents. The imidazole and phenol groups will both couple well with diazonium salts, and the imidazole group will, in addition, often react with various alkylating agents.

A. *Amine Site Reactions*

A wide variety of chemical reactions can occur at the reactive amino groups present in protein molecules. Acylating groups such as acyl halides are effective.

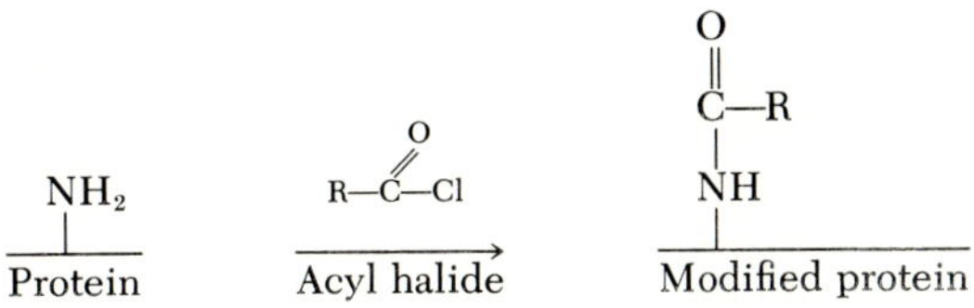

Acyl azides can modify the amino group as follows:

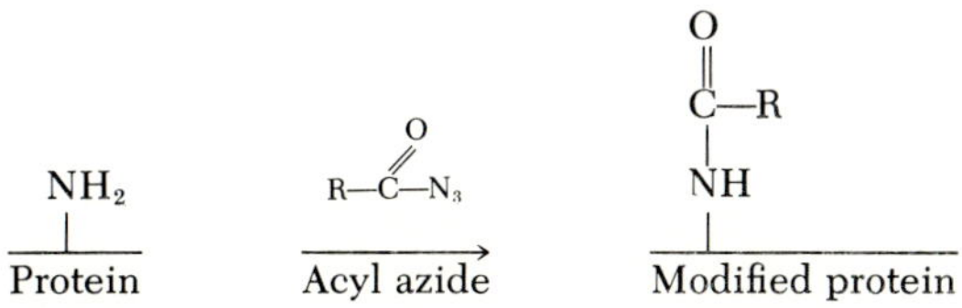

Reactions utilizing acid anhydrides go smoothly.

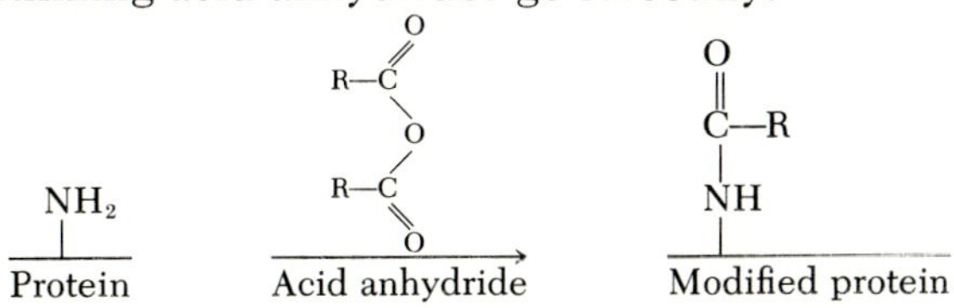

Carboxyl-containing compounds activated by carbodiimide compounds will form derivatives with proteins.

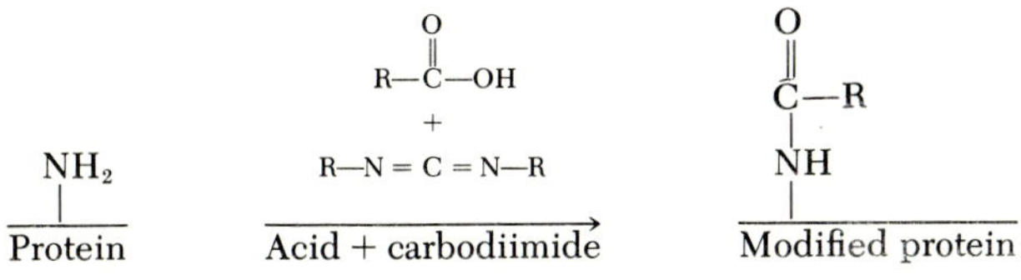

Reaction between sulfonyl chloride compounds and proteins can also result in the following product:

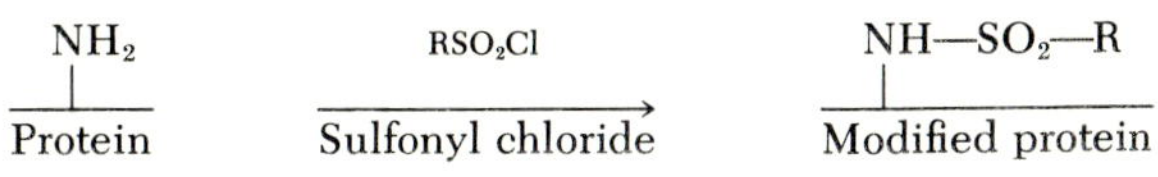

Anderson in 1954 demonstrated that isocyanates react primarily with protein amino groups under mild conditions to yield the corresponding ureido derivatives.

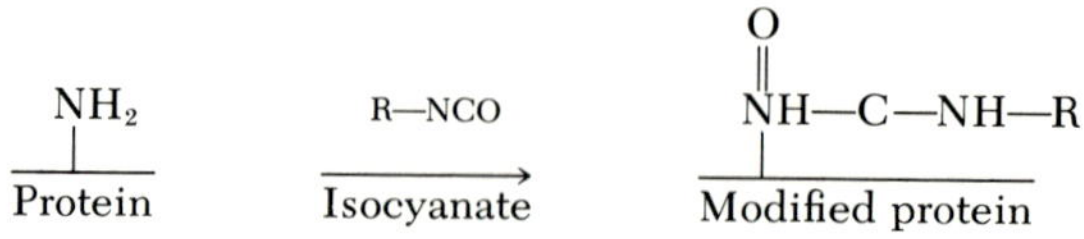

Materials of this nature have been under investigation for use as surgical adhesives and recent reports by Cooper *et al.* (1967) have detailed some of their properties. Table I illustrates the strong adhesive bonds obtained using products of this type.

Isothiocyanates are also known to react with amino groups to yield the corresponding thioureido derivatives. These materials have also been investigated as protein binding materials and not as surgical adhesives.

TABLE I

ADHESIVE STRENGTH OF SOME SURGICAL ADHESIVE SYSTEMS

Adhesive	Bond strength (g)[a] After 1 hr	Bond strength (g)[a] After 18 hr in Ringers solution
Eastman 910	200	200
Polymeric isocyanate	1200-1800	1200-1500
Gelatin/resorcinol/formaldehyde	500	900

[a]Measured on ½-inch diameter plugs of fresh lean beef.

B. Reactions at Other Sites of a Protein Molecule

Table I also lists a gelatin/resorcinol/formaldehyde surgical adhesive. This material was developed at Battelle and has already been the object of numerous reports by Cooper and Falb (1966, 1968) and Braunwald *et al.* (1966; Braunwald and Tatooles, 1966a,b). Basically, the system used formaldehyde to cross-link gelatin with living proteinaceous tissues. Resorcinol also interacts in the system and provides good water resistance to the final product.

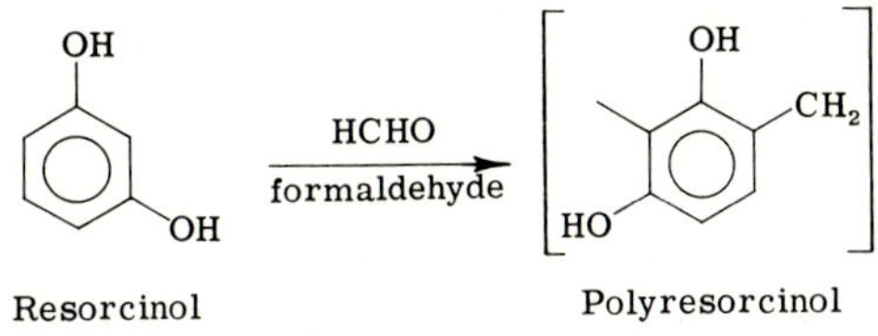

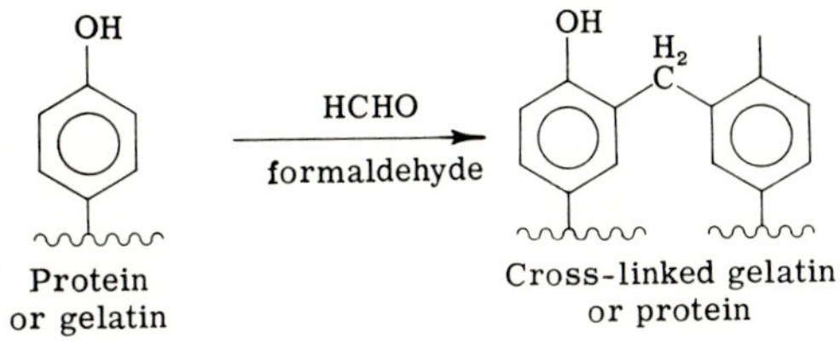

The above scheme is a somewhat simplified version of the actual chemical events; however, the equations do illustrate another possibility for tissue adhesive bonds.

Campbell *et al.* (1951), Manecke *et al.* (1960), Manecke and Singer (1960a,b), Manecke (1962), and Patchornik (1962) describe the use of alkylating and acylating reagents for binding proteins. Iodoacetic acid and dinitrofluoro benzene react under mild conditions not only with the α-amino, ϵ-amino, and thiol groups of proteins, but also with

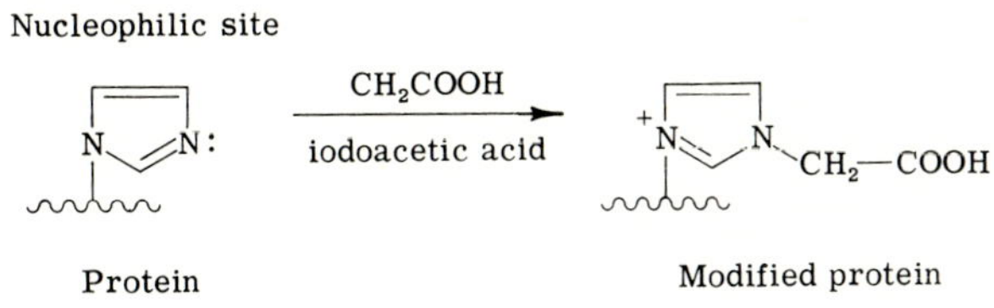

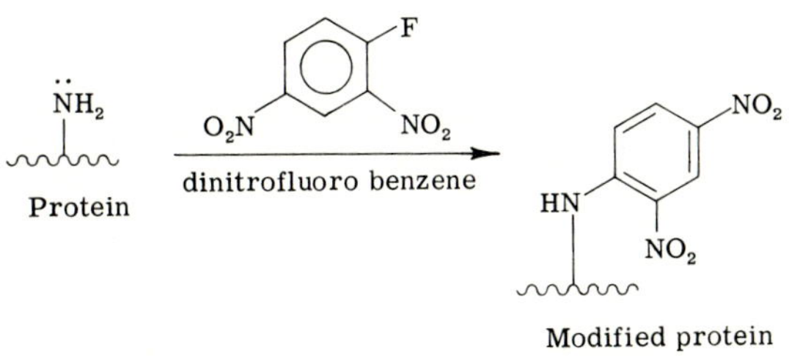

the sulfur atom of methionine, the phenolic hydroxyl of tyrosine, and the imidazole of histidine.

The coupling reaction between a polydiazonium salt and a protein is extremely rapid even at low temperatures and can be performed in neutral aqueous solutions. Moreover, the stable aromatic amine precursors may be quantitatively diazotized readily and simply before use.

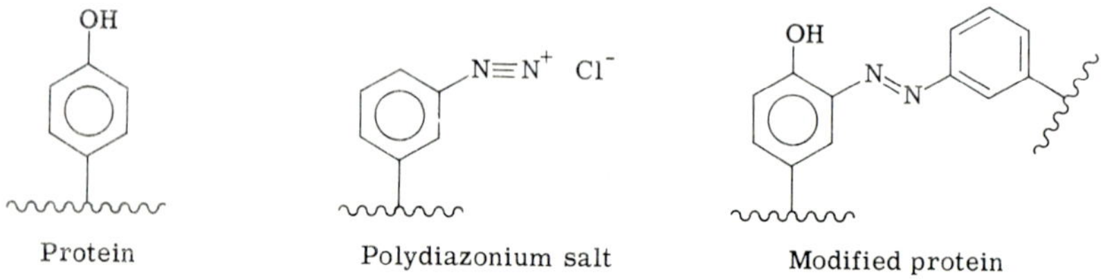

The first report on a material of this sort was by Campbell *et al.* These workers were successful in coupling bovine serum albumin with diazotized *p*-aminobenzyl cellulose. Since that time, many studies have appeared in which diazonium salts of suitable polymers were utilized to prepare active insoluble protein derivatives. No one to the writers' knowledge has investigated the potentials of such products for use as surgical adhesives.

C. *Specific Adhesive Systems*

Thus far much has been said concerning the functional groups which might be used to provide strong adhesion through primary chemical bonding. In considering specific adhesives, the desirable physical attributes should be kept in mind. Good inherent mechanical strength, solubility in aqueous environment, and hydrophilic or hydrophobic nature are all important to the final attributes of a specific adhesive system. Most of the desirable characteristics can be built into polymeric or prepolymer moieties containing multifunctional groups designed to react and cross-link with proteinaceous materials. It is not strictly necessary for the prepolymeric adhesive to be soluble or even highly fluid in nature if some effective means can be devised for its distribution between tissue sites. Highly divided powdered solids might interact with solid tissues at the interface, providing strong adhesive bonds.

Adhesives containing acylating groups capable of strong adherence with tissue via amide links might be considered. These would include acid chlorides of carboxylic or sulfonic acid ion-exchange resins.

Finely powdered carboxymethyl cellulose azide might be envisioned as a highly reactive surgical adhesive with the potential of strong adherence. Carboxymethyl cellulose itself activated by dicyclohexyl carbodiimide has been used to bind proteins. This system may be somewhat slow to harden. Investigations of systems of this type for surgical adhesive applications have not been carried out, and little is known concerning the actual reaction rates which might be achieved under optimum conditions.

Ethylene maleic anhydride copolymers can be used as acylating agents for proteins. Amino groups of protein react with the anhydride groups of polymer to form amide links. These materials should be investigated as possible surgical adhesives.

Primary chemical bonds can be formed by alkylation of the proteins with bromoacetyl cellulose derivatives. A copolymer of methacrylic acid and methacrylic acid 3-fluoro-4,6-dinitro-anilide can be used to bind proteins covalently by an acylation reaction.

Polydiazonium salts have been used as discussed previously to bind proteins and certainly appear to have promise as surgical adhesive materials. These include diazonium derivatives of cellulose (diazotized forms of *p*-aminobenzyl cellulose and the *m*-aminobenzyloxymethyl ether of cellulose) and diazotized poly-*p*-aminostyrene and also that obtained by the diazotization of a synthetic copolymer of *p*-aminophenylalanine and leucine.

The polymers containing multiisocyanate groups investigated in this laboratory as surgical adhesives appear to be quite promising. Poly-*p*-aminostyrene can be converted to the isothiocyanate derivative by treatment with thiophosgene, and this material has been used to bind proteins.

Thiophosgenation of amino derivatives of Sephadex have also led to successful protein binding agents. These products should be good surgical adhesives since they will not react with environmental moisture as easily as do the isocyanates.

Another fruitful area of investigation would probably include bi- or multifunctional reagents used to cross-link proteins. The characteristics of some materials have been described in detail by Richards (1963), Zahn (1964), and Wold (1967).

Some of these cross-linking agents are illustrated in the following:

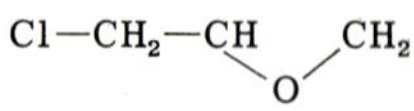

Epichlorohydrin

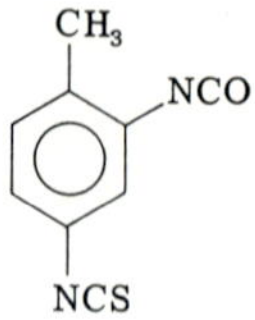

Toluene-2-isocyanate-4-isothiocyanate

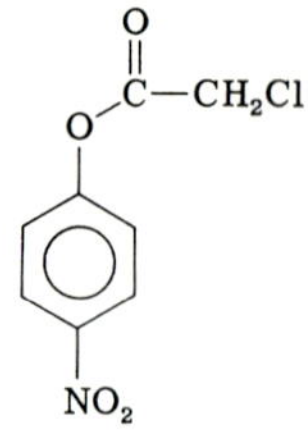

p-Nitrophenyl chloroacetate

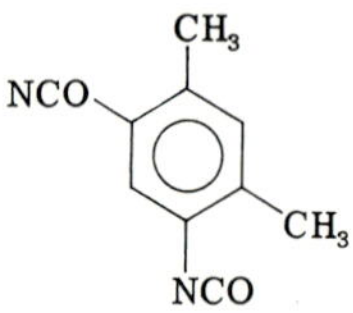

m-Xylylene diisocyanate

$C_6H_{11}—N{=}N—C_6H_{11}$

bis-Diazohexane

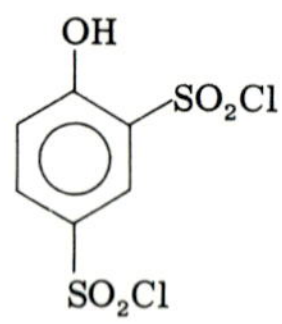

Phenol-2, 4-disulfonyl chloride

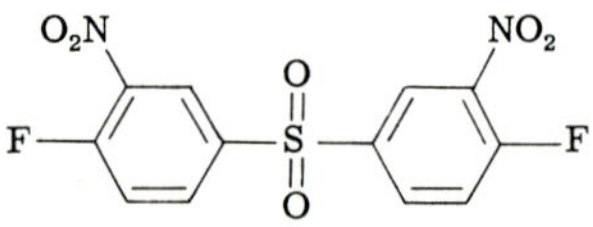

p, *p*′-Difluoro-*m*, *m*′-dinitrodiphenylsulfone

These compounds have two potential binding sites that will react with a protein molecule and should therefore be suitable for surgical adhesives.

In conclusion, a number of materials have been presented which show potentially useful applications as surgical adhesives. Major innovations may be necessary to adapt specific systems to this use, but some of the systems and approaches outlined here may answer the need for high strength adhesive materials.

References

Anderson, W. (1954). *Acta Chem. Scand.* **8**, 1723.
Braunwald, N. S., and Tatooles, C. J. (1966a). *Surg. Forum* **16**, 345-346.
Braunwald, N. S., and Tatooles, C. J. (1966b). *Surgery* **60**, 857-861.
Braunwald, N. S., Gay, W., and Tatooles, C. J. (1966). *Surgery* **59**, 1024-1030.
Campbell, D. H., Luescher, E., and Lerman, L. S. (1951). *Proc. Natl. Acad. Sci.* **37**, 575.
Cooper, C. W., and Falb, R. D. (1966). *New Scientist* **30**, 308-309.
Cooper, C. W., and Falb, R. D. (1968). *Ann. N. Y. Acad. Sci.* **146**, Art. 1, 214-224.
Cooper, C. W., Falb, R. D., and Grode, G. A. (1967). *Preprints, Am. Chem. Soc. Natl. Meeting, Chicago, 1967*, pp. 125-128.
Katchalski, E., and Silman, H. (1966). *Ann. Rev. Biochem.* **33**, Part II.
Manecke, G. (1962). *Pure Appl. Chem.* **4**, 507.
Manecke, G., and Singer, S. (1960a). *Makromol. Chem.* **37**, 119.
Manecke, G., and Singer, S. (1960b). *Makromol. Chem.* **39**, 14.
Manecke, G., Singer, S., and Gilbert, K. E. (1960). *Naturwissenschaften* **47**, 63.
Patchornik, A. (1962).
Patchornik, A. (1962). Israel Patent 18207.
Richards, F. M. (1963). *Ann. Rev. Biochem.* **32**, 269.
Wold, F. (1967). *Methods Enzymol.* **11**, 617.
Zahn, H. (1964). *Proc. 6th Intern. Congr. Biochem., New York, 1964*, Abstr. II, p. 135.

CHAPTER 10

COMPOSITIONS PRODUCING ADHESION THROUGH HYDRATION

James L. Chen
and
Gilman N. Cyr

I. Development of Adhesives Dependent on Hydration

In 1957, when the Squibb Institute for Medical Research initiated a modest program to develop adhesives for biological systems, it was recognized that attempts to treat lesions of the oral mucous membranes were hampered by the difficulty in maintaining medications at the site of application. A desirable development apparently would be a protective mechanical covering which could withstand salivation, tongue movement, and swallowing. Scrivener and Schantz (1947) had experimented successfully with gum tragacanth and commercial denture adhesive powders as carriers for penicillin in the treatment of Vincent's infection. The dry dispersions of penicillin in the powders were applied by means of a powder blower over the infected areas. Rothner and his associates (1949) had reported that carboxymethylcellulose in petrolatum represented an improved vehicle for the oral application of penicillin, providing longer duration of contact, but there was no further report of these uses.

Various hydrocolloids were initially explored. These are represented by many vegetable gums and animal proteins known to produce "tack" or to become sticky when moistened with or hydrated by water. Because ointments or pastes were considered to be generally more useful and more familiar vehicles for drugs for topical application, the hydrocolloids were dispersed in oleaginous vehicles to prepare adhesive ointments. Objective methods of measuring the adhesiveness of these preparations to wet tissue were not available, but it was possible to observe differences in the viscosity of pastes prepared from various gums. For example, pastes made from comparable concentrations of sodium carboxymethylcellulose were much more viscid and difficult to spread than pastes made from pectin or gelatin. Mixtures of gums were evaluated and a composition was selected for clinical study, consisting of equal parts of very finely ground pectin, gelatin, and sodium carboxymethylcellulose (Orahesive* powder) incorporated at a level of 50% in a polyethylene and mineral oil gel base. The first clinical evaluation of this composition, known as Orabase,* was reported by Kutscher and his associates in 1959. The average duration of maintenance at the site varied from 15 minutes to 2½ hours, depending on the location of the application and on the amount of vehicle used.

In the meantime, Kanig and Manago-Ulgado (1965) developed *in vitro* test procedures to make comparative evaluations of various dental adhesive ointments. They designed an apparatus to show the capacity of various denture adhesive ointments to absorb water through a dialyzing cellophane membrane. Water absorption for each denture adhesive ointment, including Orabase, was plotted against time. The wet adhesive strength develops as the hydrocolloid particles pick up water; it reaches a peak and then weakens as excessive

*Orabase and Orahesive are registered trademarks of E. R. Squibb & Sons, Inc.

water is absorbed. The endpoint is the time required to pull the slides apart when the separating force was 250 gm.

Early clinical experience with the adhesive ointment indicated that a powder vehicle would be desirable, and studies were undertaken on the finely powdered hydrocolloids alone, that is, a mixture of equal parts of pectin, gelatin, and sodium carboxymethylcellulose. Adhesive pastes and powders offer useful properties; but although they are capable of maintaining prolonged contact of medication with wet membranes, one surface of the application is still exposed to the erosive action of saliva. It appeared that an adhesive bandage with a flexible, water impervious backing should retard the erosion of the adhesive. Various types of bandages were formulated, but the type found to be most desirable for further development consisted of Orahesive powder dispersed in polyisobutylene, rolled into sheets and laminated to thin polyethylene film.

The bandage adheres to either wet or dry surfaces. Adherence to a dry surface is probably through mechanisms similar to those for pressure sensitive adhesives, because of the polyisobutylene content. Adherence to wet surfaces, on the other hand, is undoubtedly dependent on the hydrocolloid, but the mechanism is not clear. Since adhesion to a wet surface such as gingiva or teeth is of primary concern in the development of an intraoral bandage, information on the mechanism of wet adhesion was sought. The behavior of various hydrocolloids were compared in intraoral bandage for adherence to wet surfaces. Various test methods were devised and applied to scores of hydrocolloids in an attempt to correlate the physicochemical properties of those hydrocolloids with adhesive performance.

II. Measurements of Adhesion Through Hydration

Salter (1965) generalized that it is impossible to assess the quality of an adhesive by any simple numerical test of bond strength, and implied that it is virtually impossible to reproduce the actual mode of failure of an adhesive in service. Huntsberger (1967) stated that efforts to arrive at meaningful values for adhesion through experimental data cannot be successful. The inability to measure the interfacial interaction between solid and viscoelastic materials makes it very difficult to develop a theory of adhesion. Therefore, the test methods which were devised in this study are meant to give comparative measures of performance rather than to measure absolute adhesive bond strengths of hydrated hydrocolloids.

Several factors might be involved in the adhesive failure of intraoral bandage. One group would be the characteristics of the adhesive itself, such as adhesive modulus (stiffness), the percentage and kind of adhesive component in the mass, and the presence of air bubbles or other adhesive blocking substances. The second group would be the characteristics of the substrate, that is, its chemical nature, surface condition, and degree of wetness. The third group would be the factors related to the forces employed, as for example, the rate of shear or peel, the angle of separating force, and the contact area of the bandage and substrate. Wherever possible, adequate control of these factors was maintained in all testing.

A. *In Vivo* Screening Test

First, a simple *in vivo* screening test was designed for overall performance of intraoral bandages. The bandages were prepared by mixing the test hydrocolloid powder (60%) with polyisobutylene (40%), pressing this mass into discs about 1/16 inch thick, and covering one side with a thin polyethylene film of equal size. The exposed side was then pressed with the finger onto the anterior gingiva for half a minute. If the disc did not stick, pressure was continued for half a minute longer. All candidate compositions were first screened by this simple test to determine relative performance in a semiquantitative manner.

An excellent score (E) referred to a composition which contained 60% or less of hydrocolloid, adhered under normal application pressure within 30 seconds, and was difficult to dislodge. "Satisfactory" (S) described the same performance except that the adhesive was more easily dislodged. The bandages designated as "Fair" (F) required more pressure and/or time to adhere and were even more easily dislodged. "Poor" (P) samples usually took longer than a few minutes to develop adhesiveness, even under pressure, and were the easiest of all to dislodge, even up to 65% hydrocolloid content.

Approximately 100 individual hydrocolloids, including various grades or types of the same hydrocolloid, have been screened by this method. Table I shows the results obtained on a partial list of hydrocolloids. Mixtures of hydrocolloids have also been evaluated by this method.

B. Simulation of Intraoral Conditions

Failure of the bandage to adhere intraorally seems to result from tongue action and disruption of the gel in saliva. The apparatus shown in Fig. 1 was designed to simulate tongue action. It consists of a cylin-

TABLE I
ADHESIVE PROPERTIES OF VARIOUS HYDROCOLLOIDS IN POLYISOBUTYLENE AS MEASURED BY *in Vivo* SCREENING TEST (PARTIAL LIST)

Hydrocolloid[a]	Adhesive property[b]
Acacia	P
Agar agar	P
Alginates	
Alginic acid (Kelacid, Kelco)	P
Sodium alginate (Kelcosol, Kelco)	E
Carboxymethylcellulose, sodium	
(Hercules Type 4)	P
(Hercules Type 7, 9, 12)	E
Carboxymethylcellulose, calcium(Robeco-C-70)	P
Dextran (Industrial L,H,HH,P,2P, Pharmachem)	P
Gelatin U.S.P.	F
Guar gum (Jaguar, Stein Hall)	
Jaguar A-20-D 60%	S
Jaguar A-20-D 65%	E
Hydroxyethylcellulose	
(Natrosol 250J, Hercules)	E
Karaya gum 60%	S
Karaya gum 65%	E
Methylcellulose (Methocel, Dow)	
Methocel MC, 10 cps	E
Methocel MC, 15 cps	S
Methocel MC, 100 cps, 1500 cps, 4000 cps	P
Pectin, N.F. 60%, 65%	P
Polyethylene glycol (Union Carbide)	
Carbowax 20M	P
Polyox WSR 35	P
Polyox WSR 301, N-750, N-3000	E
Polyvinylpyrrolidone (K15, K30, K90, Antara)	P
Retene 205L, 205MS	
(Cationic water soluble polymer, Hercules)	E
Tragacanth	E

[a]Hydrocolloid present at 60% level unless specified otherwise.
[b]See text for description of values assigned to adhesive property.

der lined with wet dialyzing cellophane. The cylinder oscillates once per second inside a concentric lucite ring that presses a soft polyurethane foam ring against the cylinder. The assembly was kept at 37°C.

The bandages, cut into ½ inch squares or discs of ½ inch diameter, were applied to the wet cellophane membranes in two rows, the control samples and the test samples being alternated within and between rows. The control samples were intraoral bandages consisting of 60% Orahesive powder and 40% of polyisobutylene with polyethylene film backing of 0.75 mm thickness.

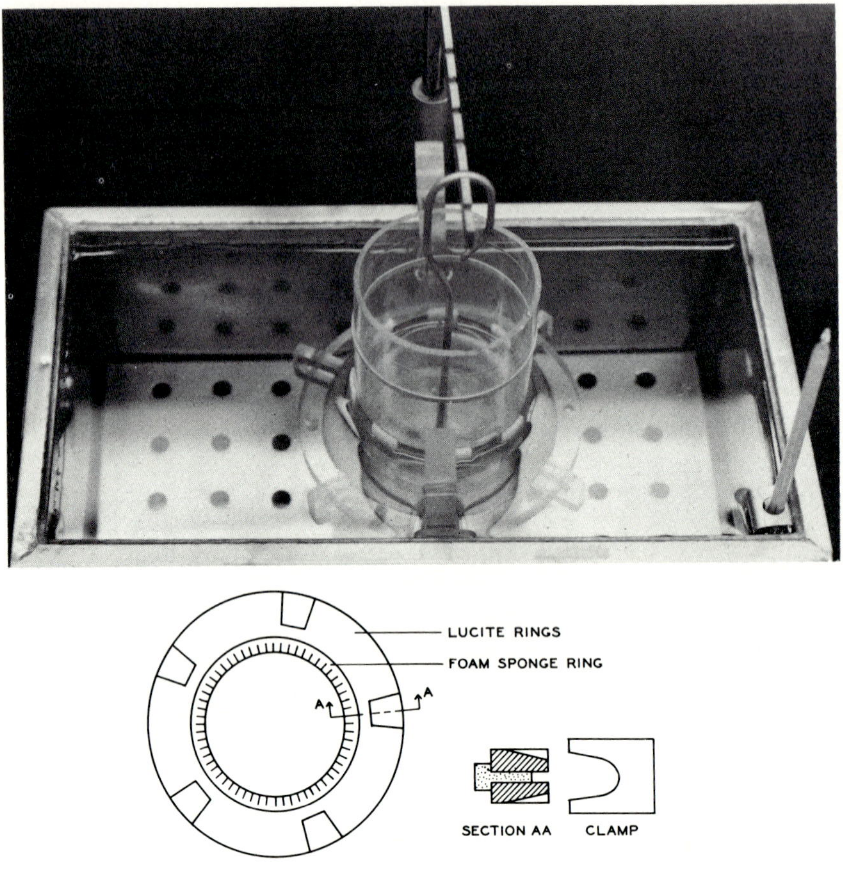

Fig. 1. Mouth action simulation apparatus.

This test can be used for the comparative evaluation of various backing materials, various shapes and sizes of bandages, and relative resistance to erosion of various bandage masses, but it appears to measure a number of different interdependent attributes of a bandage, making accurate quantitation difficult.

C. Wet Adhesive Strength

Two *in vitro* test methods have been developed to provide a quantitative expression of wet adhesive strength by measuring the force necessary to separate a sample bandage from a wet substrate. In both tests a sample bandage is pressed against wet dialyzing cellophane while the backing side of the bandage is cemented to a plastic slide by means of double-faced tape.

1. TENSILE TESTS: LAP-SHEAR AND BENDING

The test is an adaptation of those described by Mutimer *et al.* (1956) and by Kanig and Manago-Ulgado (1965). The apparatus is shown in Fig. 2. The lap-shear measurements are made with the direction of the pull in line with the cement layer. For measuring the bending resistance, the slides are placed perpendicular to the base of the platform; depending on the slide placement in relation to the height of the pulley, various angles of pull may be studied. In each case, incremental weights are gently placed on the pan until the adhesive bond is broken.

The lap-shear tests appear to correlate with the results of the *in vivo* screening tests in many cases, as shown in Table II. Since the *in vivo* screening test measures total adhesive performance as affected by pushing, pulling, and angular lifting forces, the adhesive performance of a bandage sometimes cannot be judged by one test alone. The *in vitro* tests usually reflect behavior under single forces. For example, three methylcellulose bandages, MC 25 cps, MC 1500 cps, HG 50 cps, rated poorly by the *in vivo* screening test, gave good results (150-250) with the lap-shear test, but with the 70° angle tangential test these bandages gave poor results (< 20). In all, over 100 hydrocolloids have been studied, either alone or in combination with other substances. A similar procedure has been applied to measure the force required to cause separation when a movable platform carries a Chatillon strain gauge (Fig. 3).

2. *In Vivo* PEEL ADHESION TEST

A Chatillon gauge was also used to measure the force required to separate the bandage from oral mucous membrane or from the teeth. As shown in Table III, the peel adhesion tests agree reasonably well with those obtained in the *in vitro* tests in spite of the fact that the curvature of the gingiva is variable, manually applied pressure is not controllable, and rate of pull is also uncontrollable. The adhesive bond is stronger on the teeth than it is on soft tissue or wet cellophane, perhaps because greater maximum pressure can be applied.

D. In Vitro Test for Duration of Adhesion

The lap-shear test does not measure adhesive strength in an environment resembling the oral cavity. Such an environment is repro-

Fig. 2. Tensile apparatus shown in position for lap-shear test. Inset shows similar set-up for bending test.

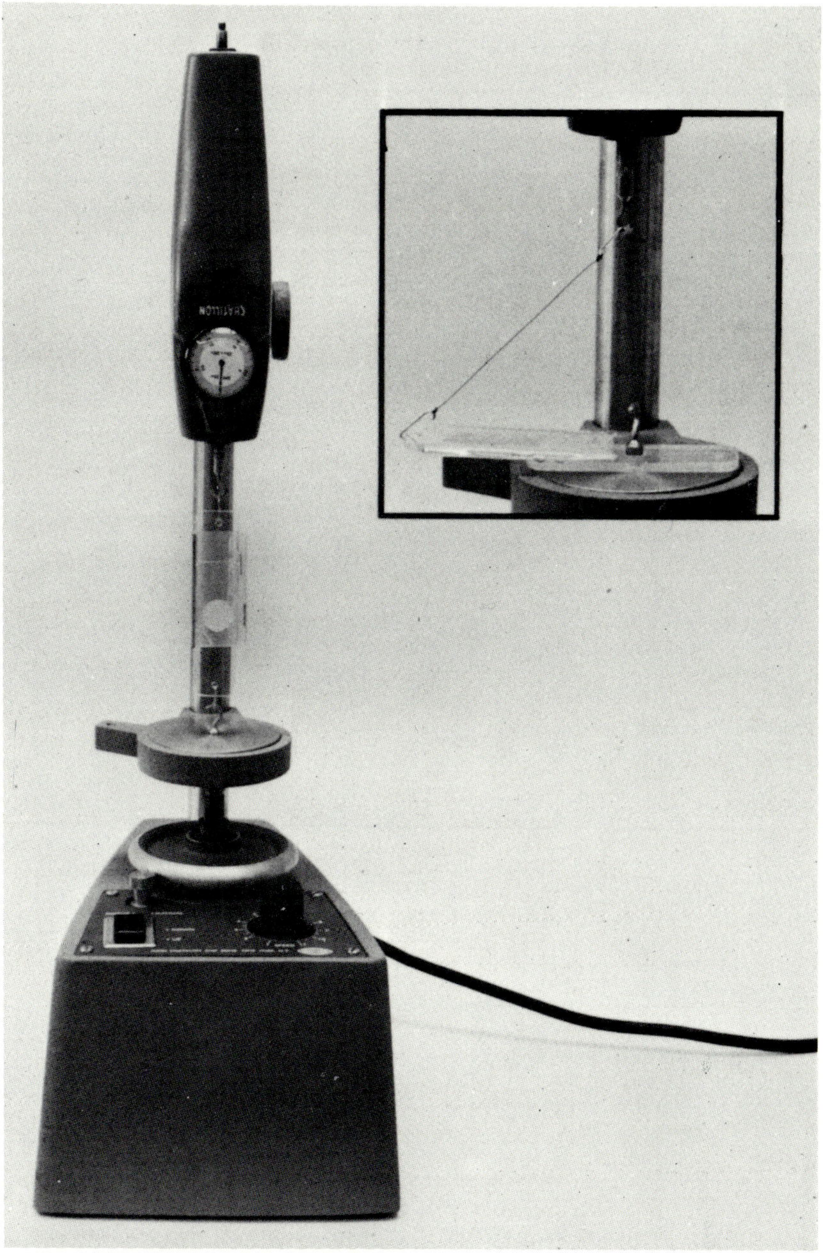

Fig. 3. Chatillon Gauge shown in position for lap-shear test. Inset shows similar set-up for bending test.

TABLE II
COMPARISON OF RESULTS OBTAINED ON BANDAGES BY *in Vivo* SCREENING VS. LAP-SHEAR TESTS (PARTIAL LIST)

Intraoral bandage containing 60% of listed hydrocolloid	*In vivo* screening test: Adhesive property	Lap-shear test: Weight (gm) required to break	
		Shearing pull	70° angle Tangential pull
Dextran Indust. HH, Pharmachem	P	<7-20	20-40
Dextran Indust. P, Pharmachem	P	<7	–
Polyox WSR35, Union Carbide	P	20-25	–
Dextran 2P, Pharmachem	F	20-50	60-70
Guar gum (Jaguar A-40-F, Stein Hall)	S	350-375	–
Karaya gum	S	50-100	–
Control (Orahesive bandage)	E	150-200	200-250
Hydroxyethylcellulose (Natrosol 250M, Hercules)	E	150-170	100-120
Methylcellulose (Methocel MC, 10 cps Dow)	E	150	50-80
Sodium carboxymethylcellulose, type 7H, Hercules	E	170-200	100-150
Retene 205MS, Hercules	E	200	150-270

TABLE III
COMPARISON OF *in Vivo* AND *in Vitro* RESULTS OBTAINED WITH CHATILLON GAUGE USING ORAHESIVE BANDAGE

Substrate	Gauge readings (gm)
In vivo	
Lower inner lip	90-360
Lower gingiva	130-325
Upper anterior teeth	680-950
In vitro	
Wet dialyzing cellophane	150-185

duced in part by the simulation test—a test which does not measure duration of adhesion. This property is measured while the bandage is constantly bathed in water and subjected to a constant strain. The time necessary for a given force to break the adhesive bond is determined while the bandage is immersed in a circulating water bath maintained at 35°–37°C. The apparatus is shown in Fig. 4.

Table IV summarizes the results of the duration of adhesion test on bandages prepared from different organic hydrocolloids. The bandages were allowed to "hydrate" for 20 minutes in contact with the wet cellophane before being immersed in a water bath for testing. The tests were run with a force of 50 gm on the samples immediately after they were immersed in the bath. The duration of adhesion depends on force as shown in Table V.

The test for *in vitro* duration of adhesion differs in one significant respect from the situation which prevails in the oral cavity. During *in vitro* testing, additional water permeates through the wet cellophane because the test piece is completely submerged in water. In the *in vivo* situation, the saliva is limited to that which lies on the mucosal surface; additional fluid is available only to the edges of the bandage. The free passage of water in the *in vitro* test causes further and more

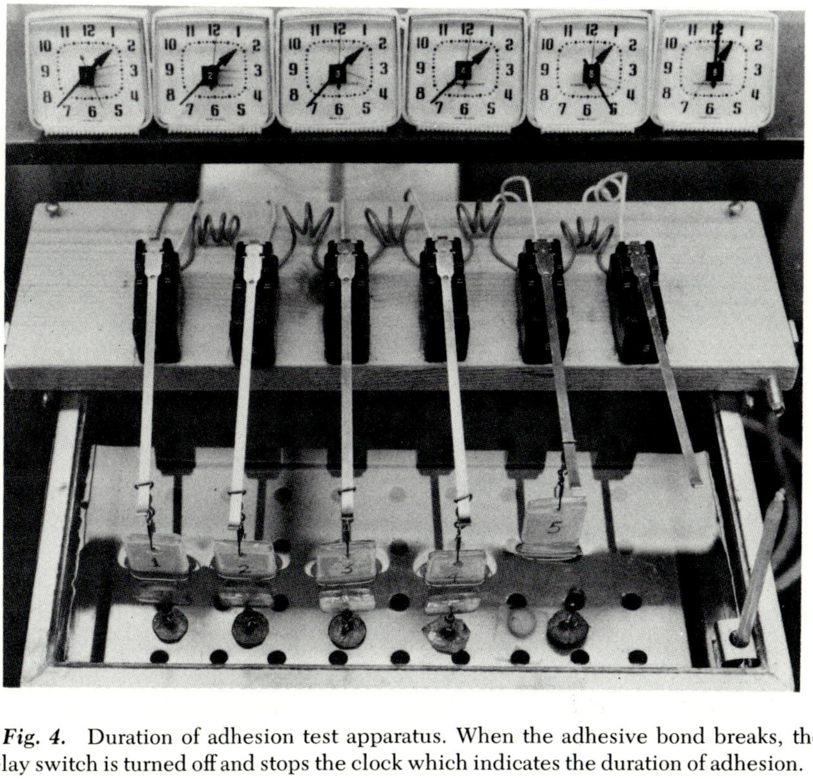

Fig. 4. Duration of adhesion test apparatus. When the adhesive bond breaks, the relay switch is turned off and stops the clock which indicates the duration of adhesion.

TABLE IV
In Vitro DURATION OF ADHESION OF INTRAORAL BANDAGES MADE FROM VARIOUS HYDROCOLLOIDS AND POLYISOBUTYLENE

Sample[a]		Duration of adhesion[b]
Control (Orahesive bandage)		3.8–6 min
Sod. CMC	40%	5.5 min
Gelatin	17.5%	
Pectin	2.5%	
Sod. CMC	20%	3.3 min
Gelatin	20%	
Jaguar A-40-F	20%	
Sod. CMC	60%	2.4 min
Jaguar A-40-F	10%	
Jaguar A-20-D	70%	3.3 hr
Natrosol 250 J,	63%	1.7 min
Retene 205 MS	60%	9.3 min

[a]Sample consists of hydrocolloids incorporated at specified levels in polyisobutylene.
[b]For 0.2 inch square bandage under a 50 gm loading.

TABLE V
EFFECT OF FORCE ON DURATION OF ADHESION OF ORAHESIVE BANDAGE

Weight of pull (gm)	Time (min) required to break[a]
25	15.5–25.3
50	10.7–18
100	1–4.5

[a]Bandages were allowed to "hydrate" for 20 minutes in contact with the wet dialyzing cellophane before being immersed in water.

rapid hydration of the hydrocolloid particles. As this additional water is absorbed the adhesive bond is weakened owing to the formation of a slippery mucilage. Consequently, the *in vitro* test may very well be considered to be an accelerated test for duration of adhesion. Some evidence of this is seen in comparing Orahesive bandage under both conditions. By the *in vitro* test, Orahesive bandage showed a duration of adhesion ranging from 3.8 to 25 minutes, depending on the weight of pull and hydration time. On the other hand, under actual clinical conditions Orahesive bandage has a duration of adhesion ranging from 6 to 24 hours.

III. Discussion of Wet Adhesive Test Methods

All adhesive test methods are affected by such conditions as weight and angle of pull, rate of shear, size and type of contact area, temperature, modulus, and viscoelasticity. In the wet adhesive test methods just described these factors are controlled insofar as possible. Additional factors were found to cause variations in test results.

A. The Substrate Test Site

Hog casing has been tried as an alternate substrate in lap-shear test but was discarded because the results were often variable from one site to another, and only reproducible when tested on the same site. Dialyzing cellophane was selected as a more convenient substrate. The surfaces appeared to be quite smooth and homogeneous but variations in results were still obtained, as can be seen in Table VI. To obtain reproducible results, the test sites should be carefully chosen. The sites must be those which give a normal result with the control bandage and which can be washed and reused in all subsequent tests.

B. Effect of the Amount of Water on Adhesive Strength

The duration of adhesion depends on the amount of water at the interface. When a dry disc of Orahesive bandage is applied with pressure to wet dialyzing cellophane, and the bandage allowed to soak up

TABLE VI
VARIANCE AMONG TEST SITES ON CELLOPHANE

Dialyzing cellophane test site no.	Weight (gm) required to break[a]
1	190
2	190
3	170,180
4	120,120
5	210
6	130, 100, 140
7	180
8	200

[a]Orahesive bandage used as test material.

TABLE VII
DURATION OF ADHESION OF ORAHESIVE BANDAGE AS AFFECTED BY VARYING AMOUNTS OF WATER AVAILABLE FOR HYDRATION

Water available for hydration (ml/sq in)	Duration of adhesion (min) (50 gm load)
0.013	3.8–6[a]
0.025	3
0.050	1.3

[a]Data reported in Table IV.

the water at the interface but removed before adhesive forces become established, the bandage disc increases in weight equivalent to about 0.013 ml of water per square inch. This compares with about 0.012 ml of saliva per square inch of oral mucosa. In order to introduce more water at the interface, water was added directly to the exposed surface of a disc of dry Orahesive bandage, spread evenly, and allowed to be absorbed completely into the bandage before it was pressed onto the surface of dialyzing cellophane from which the surface water had been removed by means of blotting paper. In one case 0.025 ml of water/square inch was added to the disc and in another, 0.05 ml/square inch. The results shown in Table VII demonstrate clearly that the larger amounts of water available for hydration have a deleterious effect on adhesive forces.

The effect of the total amount of water available for hydration on the adhesive property of various other hydrocolloids is shown in Table VIII. It can be seen in all the compositions that were studied that an excessive amount of water reduces duration of adhesion. However, the magnitude of this change is not the same for each gum. The amount of water available for hydration is very difficult to control and may be the major source of error in all tests for wet adhesiveness.

C. *Hydration Time*

Wet adhesive strength and duration of adhesion are affected by the time allowed for hydration after pressing the bandage onto a wet surface to achieve initial adhesion. Table IX shows that when a bandage is tested immediately after it is applied to the substrate, some degree of adhesiveness is obtained. However, when the bandage is allowed to stand on the wet substrate for a period of time before it is tested, its adhesive is much greater.

Table X shows a similar relationship between hydration time and duration of adhesion.

TABLE VIII
DURATION OF ADHESION AS AFFECTED BY THE AMOUNT OF WATER AVAILABLE FOR HYDRATION

Sample[a]	Amount of water delivered to the sample surface: 0.005 ml	0.010 ml
	Duration of adhesion[b]	
Sodium CMC 20% Gelatin 20% Jaguar A-40-F 20%	10 min, 20 min	6.7 min, 8 min
Sodium CMC 40% Jaguar A-40-F 10% Solka Floc[c] 10%	3.7 min	6 sec
Sodium CMC 45% Jaguar A-40-F 10% Solka Floc[c] 5%	6 sec	3 sec
Natrosol 250 J 63%	1 min, 2.3 min	32 sec, 45 sec
Retene 205 MS	15 min	8 min

[a]Sample consists of hydrocolloids incorporated at specified levels in polyisobutylene.
[b]For a 0.2 inch square bandage under a 50 gm loading.
[c]Refined cellulose powder, Brown Co.

TABLE IX
THE EFFECT OF HYDRATION TIME ON WET ADHESIVE STRENGTH OF ORAHESIVE BANDAGE

Hydration time (min)	Gauge reading (gm)[a]	No. of readings
Immediate	125–170	4
5	350–375	3
15	400–575	5
20	725–1000	6

[a]Chatillon gauge modification of lap-shear test.

TABLE X
EFFECT OF HYDRATION TIME ON DURATION OF ADHESION OF ORAHESIVE BANDAGE

Hydration time (min)	Time (min) required to break (50 gm pull)
Immediate	3.8, 6
5	6.2, 6.2
10	9.3, 10
20	10.7, 18

IV. Properties of Hydrocolloids Related to Wet Adhesion

A. *Molecular Characteristics of Wet Adhesives*

(1) The effective wet adhesives are usually macromolecular organic hydrocolloids containing numerous hydrogen-bond-forming hydrophilic functional groups such as —OH, —COOH, —SO_4H, and —NH_2. Among the van der Waals' forces, hydrogen bonding appears to play a major role in wet adhesion.

(2) Compounds capable of inducing good wet adhesion have a molecular weight of 100,000 or more. Beyond this the molecular weight per se of a hydrocolloid does not appear to determine its relative wet adhesive strength or performance. Table XI shows the molecular weight of some hydrocolloids and their *in vivo* and *in vitro* adhesive properties. Within the range of 100,000 for the carrageenans to 19.5 million for the dextrans, both excellent and poor adhesives may be found. Within a molecular type, chain length may be a determinant of adhesive strength. For example, degraded carrageenan is much less effective as an adhesive than is the higher molecular weight potassium carrageenate. In the case of polyethylene glycols, Carbowax 20M with a molecular weight of 20,000 has no wet adhesive property. At a molecular weight of 200,000 (as for Polyox WSR 35) adhesiveness is improved, although not as much as seen at 4,000,000 (as for Polyox WSR 301) which is an excellent adhesive.

B. *Physicochemical Properties of Wet Adhesives*

1. Water Dispersibility

The formation of a colloidal dispersion in water appears to be neces-

TABLE XI
Relationship between Molecular Weight of Hydrocolloids and Wet Adhesive Quality

Hydrocolloid	Molecular weight	*In vivo* screening test: Adhesive property	Lap-shear Test: Weight (gm) required to break
Carrageenate, potassium	100,000-800,000	E	300-350
Degraded carrageenan	20,000-30,000	P	< 7
Guar gum	200,000	E	150-375
Amylopectin	162,000-800,000	P	< 7
Karaya	9,500,000	E	50-100
Dextran	17,700-19,500,000	P	< 7- 50
Polyox WSR 35	200,000	P	20- 25
Polyox WSR 301	4,000,000	E	170-200

sary in wet adhesion in order to expose the active sites of the hydrocolloid. The addition of a cross-linking agent (such as calcium) to anionic hydrocolloids [such as sodium carboxymethylcellulose (CMC)] reduces solubility and decreases the wet adhesive property of the hydrocolloid. Similarly, radiation conversion of water soluble Polyox to a highly cross-linked and water insoluble compound, Polyox I, results in the loss of its wet adhesive property. All types of sodium CMC exhibit excellent wet adhesive property except for Type 4 which has a lower degree of substitution and is, therefore, practically insoluble in water. Water insoluble macromolecules, even those containing multiple hydrophilic functional groups, such as cellulose and zein, are ineffective as wet adhesives. In these molecules, free adhesive bonds are not available for attachment to substrates because they are internally cross-linked as indicated by their high tensile strength.

2. Degree of Hydration

Hydrocolloids develop their wet adhesive property at various degrees of hydration and reach a maximum adhesion at an optimum degree of hydration. Some hydrocolloids exhibit wet adhesiveness in the presence of only very little water, whereas excessive water will cause the formation of a slippery, nonadhesive mucilage. Some of these hydrocolloids, such as polyvinylpyrrolidone and dextran, do not adhere well on the wet mucosa. This probably results from the amount of saliva on the mucosa which is greatly in excess of the maximum required to achieve an optimum hydration.

3. Ionization of Hydrocolloids

Good wet adhesives may be either anionic, cationic, or nonionic. Anionic sodium carboxymethylcellulose, cationic Retene 250, and nonionic guar gum are all excellent wet adhesives.

4. Rate and Capacity of Water Absorption

The rate and capacity of water absorbency of hydrocolloids affect the amount of water present in the vicinity of the interface between the bandage and the substrate. These properties appear to be important in determining the time required for initial wet adhesion, the hydration time, and the duration of adhesion. It is believed that the faster the water is absorbed by the hydrocolloid, the shorter is the time required for the hydrocolloid to obtain initial adhesion and maximum adhesive strength. On the other hand, rapid water absorbency may cause the shortening of the duration of adhesion because erosion proceeds more rapidly.

V. Mechanism of Wet Adhesion

Based on the limited knowledge about the relationship between molecular characteristics and physicochemical properties of hydrocolloids and wet adhesion, the following concept of the mechanism of wet adhesion may be visualized.

When the dry hydrocolloid is moistened with water, the long chains of the polymer in the dry mat are hydrated and liberated to a freely moving state. When brought close to the substrate, the stretched, entangled, or twisted molecules are now able to match their active adhesive sites with those on the hydrophilic substrate to form adhesive bonds or to match with each other (if not already well matched) to form cohesive bonds. Maximum wet adhesive strength is attained when perfect matching of active adhesive sites is achieved in the presence of an optimum amount of water at or near the interface. If insufficient water is used to hydrate the dry hydrocolloid, active wet adhesive sites are not completely liberated and exposed for interaction. An excessive amount of water, on the other hand, causes overextension of the hydrogen bonds and other adhesive forces leading to a weakening of the adhesive; bond strength decreases with an increase in the distance between molecules.

Wet adhesion is different from adhesion to dry surfaces in that it is not a static state. Water diffuses from the surface of a hydrocolloid particle into the center and from one particle to the neighboring particles until an equilibrium is reached. Furthermore, in the case of intraoral bandage, the diffusion of water or saliva into the bandage constantly changes this equilibrium.

When the hydrocolloid is adequately hydrated, it acquires a typical semisolid viscoelastic tack somewhat similar to that observed with pressure sensitive adhesives. Excessive hydration of the hydrocolloid produces a slippery mucilage or soft paste resulting in the eventual loss of the adhesiveness.

Acknowledgments

Grateful acknowledgment is made to the many co-workers who have participated in these investigations, especially Mr. W. A. Anderson, Dr. H. B. Bernstein, Dr. R. A. Heiser, Dr. A. H. Kutscher, and Dr. N. M. Ross.

References

Huntsberger, J. R. (1967). *In* "Treatise on Adhesion and Adhesives" (R. L. Patrick, ed.), Vol. 1, p. 120. Marcel Dekker, New York.

Kanig, J. L., and Manago-Ulgado, P. (1965). *J. Oral Therap. Pharmacol.* **1**, 413-420.
Kutscher, A. H., Zegarelli, E. V., Beube, F. E., Chilton, N. W., Berman, C., Mercadente, J. L., Stern, I. B., and Roland, N. (1959). *Oral Surg., Oral Med., Oral Pathol.* **12**, 1080-1089.
Mutimer, M., Riffkin, C., Hill, J., Glickman, M. E., and Cyr, G. N. (1956). *J. Am. Pharm. Assoc., Sci. Ed.* **45**, 212-218.
Rothner, J. T., Cobe, H. M., Rosenthal, S. L., and Bailin, J. (1949). *J. Dental Res.* **28**, 544.
Salter, R. (1965). *In* "Aspects of Adhesion" (D. J. Alner, ed.), Vol. 1, p. 81. Oxford Univ. Press (Univ. London), London and New York.
Scrivener, C. A., and Schantz, C. W. (1947). *J. Am. Dental Assoc.* **35**, 644-647.

APPLICATIONS OF ALPHA CYANOACRYLATES

CHAPTER 11

HEMOSTATIC APPLICATIONS OF ALPHA CYANOACRYLATES: BONDING MECHANISM AND PHYSIOLOGICAL DEGRADATION OF BONDS

Fred Leonard

I. Introduction

Bleeding resulting from local trauma, and tissue destruction is controlled by the body through a complex hemostatic mechanism. The end result of the series of reactions comprizing this mechanism is a strongly adherent, fibrin-reinforced platelet aggregate which tightly seals the bleeding vessel.

In the case of severe trauma, where blood flow is high and the body's own mechanism per se is insufficient to prevent exsanguination, man has devised means of inducing reduction in the rate of blood flow by vessel blockage and constriction. Platelet adhesion aggregation and formation of a clot are then permitted to occur in accordance with the clotting mechanism. Such methods for controlling bleeding include the use of sutures, pressure, cold, cautery, vasoconstriction, and chemicals which activate the clotting cascade at various stages such as thrombin and other products such as regenerated cellulose.

Most recently, the ability of rapidly polymerizing monomeric 2-cyanoacrylate adhesives to adhere firmly to moist tissue substrates has evoked considerable medical interest in their potential as hemostatic agents and tissue adhesives for closure of wounds in place of, or as adjuncts to, conventional surgical methods.

Certain general characteristics are desired in compounds for such applications. In addition to the usual requirements of polymers for implant use, such as nontoxicity and nonallergenicity, a tissue adhesive ideally should (a) wet and spread on the tissue substrate, (b) polymerize rapidly, (c) form a flexible polymeric film which coats the tissue substrate, and (d) be biodegradable. The first characteristic is a function of the relationship between surface energy of the substrate and surface tension of the adhesive and the interfacial tension between the adhesive and the substrate. Spreading of the adhesive is favored when the difference between the surface energy of the substrate less the sum of the surface tension of the adhesive and the interfacial tension between the substrate and the adhesive is a positive number. When the adhesive has a real affinity for the substrate, spreading will usually occur if the surface is free from contamination.

Biodegradability is an important adhesive attribute in order that the adhesive be replaceable by the body's own contiguous tissue and not serve as a barrier to healing. It is important, of course, that the metabolites resulting from the degradation of the polymer not be stored in the tissues or organs, but eliminated through the normal excretory routes.

The *n*-alkyl 2-cyanoacrylate fulfill these requirements to varying degrees; the results of investigations of these monomers as tissue adhesives and hemostasis inducing compounds, with particular regard to surface phenomena, biodegradability, and medical implicaions of the results are the subjects of this chapter.

II. Preparation and Properties of 2-Cyanoacrylates

Details for the preparative procedures for synthesizing the alkyl 2-cyanoacrylate monomers are adequately described in the patent literature (Jeremias, 1956). In general, the reaction scheme involves the reaction between a cyanoacetate and formaldehyde to form a methylol derivative from which water is split out. This reaction is accompanied by spontaneous polymerization. The polymer which forms is then cracked to monomer in the presence of sulfur dioxide which acts as a

TABLE I
PHYSICAL PROPERTIES OF CYANOACRYLATE ESTERS[a]

Cyano-acrylate ester	Molecular weight	Boiling point (°C/mm Hg)	Density[b] (g/ml)	Index of refraction[b]	Surface tension (dynes/cm)
Methyl	111	55/4	1.1044	1.4459	37.41
Ethyl	125	60/3	1.040	1.4391	34.32
n-Propyl	139	80/6	1.001	1.4408	32.80
n-Butyl	153	68/1.8	0.989	1.4424	31.11
n-Amyl	167	113/5.4	0.972	1.4440	30.25
n-Hexyl	181	90/1.6	0.958	1.4458	29.98
n-Heptyl	195	125/1.2	0.942	1.4466	30.28
n-Octyl	209	117/1.8	0.931	1.4489	29.18
Allyl	137	74/4.2	1.066	1.4586	35.38

[a]Reprinted with permission of *The Journal of Applied Polymer Science.*
[b]A + 20°C.

acid inhibitor. In addition, the monomer is usually stabilized with a free radical inhibitor (monomethyl ether of hydroquinone) to prevent inadvertent free radical polymerization.

Typical properties of *n*-alkyl cyanoacrylate monomers are summarized in Table I.

III. Polymerization Mechanism

The proposed mechanism of polymerization in accordance with Coover and co-workers (1959) is as shown in accordance with the following equation:

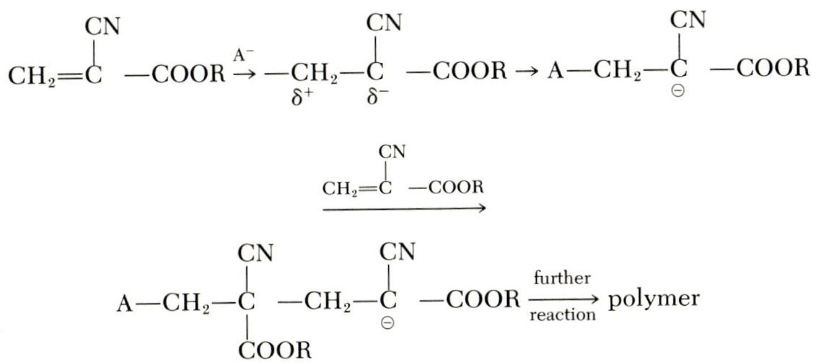

The approach of the attacking nucleophile brings about strong electromeric (—E) effects which make the nitrile and alkoxy carbonyl group highly electronegative, thereby causing polarization of the double bond. Even weak bases such as water or alcohol can apparently induce such effects and initiate polymerization. This mechanism requires that the base initiating species be present as an end group in the polymer. That such is the case may be demonstrated by the infrared data previously reported (Leonard, *et al.*, 1966). The polymer prepared using water as an initiator shows the presence of the OH group at 3600 cm^{-1}. Polymer prepared in aqueous methanol shows suppression of the band at 3600 cm^{-1} and the appearance of the OCH_3 absorption band at 1100 cm^{-1}. The data indicate an apparent competition between the methoxy and hydroxy nucleophile for initiation of polymerization. Polymers prepared in aqueous solutions of pyridine and in aqueous solutions of amino acids, such as cystine, alanine, and glycine, show substantial suppression of the OH absorption band. These data suggest that nucleophiles other than OH^- may be preferentially involved in the initiation and polymerization in the presence of amino acids such as NH_2 groups. If such is the case, the NH_2 groups of protein molecules in the tissues may possibly be involved in initiating polymerization of the monomers when used *in vivo*. Such initiation could lead to primary chemical bonding of the adhesive to the tissue substrate, thereby resulting in strong bonding. Further evidence indicating this possibility is presented by the fact that initiation of polymerization of methyl 2-cyanoacrylate with carbon-tagged glycine resulted in the production of a radioactive polymer (Kulkarni, 1967).

IV. Surface Phenomena

In the initiation of polymerization of the *n*-alkyl cyanoacrylate monomers in water, it has been observed that if the homologous series of the monomers (from methyl 2-cyanoacrylate to *n*-heptyl 2-cyanoacrylate) are dropped on distilled water in a petri dish, the rates of polymerization are approximately fastest for the lower homologs (methyl to butyl) and slowest for the higher homologs (aqueous effect). In this experiment, it is observed that the monomers from methyl to *n*-butyl spread on distilled water to a thin film and polymerize rapidly, whereas the monomers from amyl to heptyl spread on distilled water and subsequently form discrete droplets.

When monomers were used in experimental surgery to stop bleeding in surgically wounded dog tissues, by spraying the monomers against cut tissues, it was observed that the rates of polymerization were directly opposite those observed on distilled water. In this instance, the heptyl monomer was converted to a polymer film most rapidly and the methyl monomer least rapidly.

The methyl, ethyl, and *n*-propyl monomers appeared to form liquid puddles which gradually polymerized to solid polymeric, discontinuous films, whereas the higher homologs spread on the tissue surface and polymerized instantaneously (blood effect).

Average spreadability and polymerization time results for water and citrated blood illustrating these effects are summarized in Table II. Figure 1 demonstrates pictorially the spreadability obtained on blood plasma for the homologous series, illustrating a typical "blood effect." Since, in this example, the rate of interfacial vinyl polymerization may be presumed to be a function of the amount of monomer surface exposed to the substrate and the availability of the monomer to the catalyst sites present in the substrate, the higher homologs polymerize practically instantaneously and most rapidly on blood. The lower homologs which form liquid lenses and are thereby confined as bulk monomers on blood polymerize at a considerably slower rate than the higher homologs, and polymerization times of over 300 seconds are observed.

Further work showed that typical aqueous effects were obtained with physiologic saline (0.89%), 5% dextrose in water, and lactated Ringer's solution. The blood effect was obtained with dog and human blood, plasma, and serum, isotonic solution of human albumin, 1% aqueous casein solution, and 1% egg albumin. A modified blood effect (no spreadability), slow polymerization for the methyl derivative, and a small spread but rapid polymerization for the heptyl derivative, were obtained with 3% arginine solution whose surface tension had been adjusted to 35.9 dynes/cm by potassium oleate solution.

TABLE II

AVERAGE SPREADABILITY AND POLYMERIZATION TIMES FOR WATER AND CITRATED BLOOD[a]

	H_2O (pH 7.1)		Citrated blood	
Monomer	Spreadability (cm)	Polymerization (time in sec)	Spreadability (cm)	Polymerization (time in sec)
Methyl	20	10	0.5	>300
Ethyl	14.5	10	0.5	>300
n-Propyl	10	10	2	20
n-Butyl	18	10	5	1
n-Amyl	16	15	14	2
n-Hexyl	16	64	14	2.5
n-Heptyl	20	>300	11	2
n-Octyl	18	>300	15	4

[a]Reprinted with permission of *The Journal of Applied Polymer Science*.

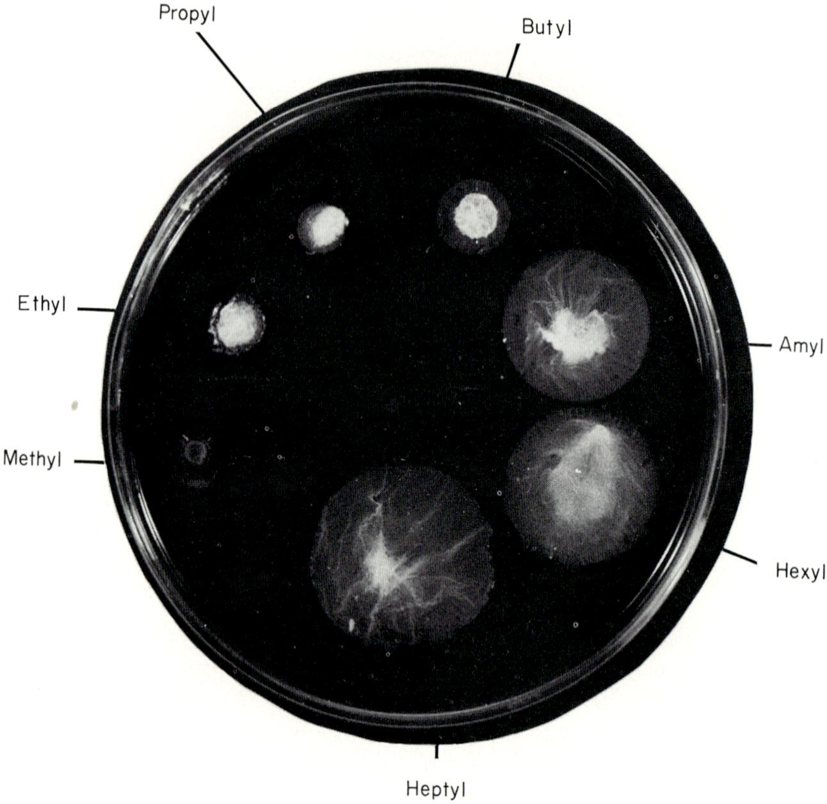

Fig. 1. Spreadability.

Interestingly enough, a 1% aqueous gelatin solution gave mixed "blood" and "aqueous" effects, e.g., large spreadability and rapid polymerization times for both the methyl and heptyl monomers. In order to study the aqueous and blood effects in more detail, spreading coefficients were determined for the monomers on both types of substrates. Attempts to measure interfacial tension of these cyanoacrylates against serum or casein solutions were not successful because rapid polymerization occurred at the interface. Instead, methyl- and heptyl cyanoacetates were substituted as model compounds for the cyanoacrylates in these instances. Pertinent data are summarized in Table III. The spreading coefficients confirm the qualitative spreading observations.

TABLE III
SPREADING COEFFICIENTS[a]

Sample A	δ_A	Sample B	δ_B	Sample AB	δ_{AB}	Spreading coefficient $\delta_A - [\delta_B + \delta_{AB}]$
		Water effect solutions				
Water/air	71.4	Methyl monomer/air	37.0	Methyl monomer/water	5.7	28.7
		Heptyl monomer/air	28.3	Heptyl monomer/water	16.6	26.5
Water/air	71.4	Methylcyanoacetate/air	42.2	Methylcyanoacetate/water	4.3	24.9
		Heptylcyanoacetate/air	26.8	Heptylcyanoacetate/water	16.9	27.7
		Blood effect solutions				
Casein/air	43.4	Methylcyanoacetate/air	42.2	Methylcyanoacetate/casein	5.7	−4.5
		Heptylcyanoacetate/air	26.8	Heptylcyanoacetate/casein	9.2	+7.4
Serum/air	45.4	Methylcyanoacetate/air	42.2	Methylcyanoacetate/serum	10.5	−7.3
		Heptylcyanoacetate/air	26.8	Heptylcyanoacetate/serum	13.0	+5.6
Arginine/air	35.9	Methyl monomer/air	37.0	Methyl monomer/arginine	6.0	−7.1
		Heptyl monomer/air	28.3	Heptyl monomer/arginine	7.4	+0.2
		Blood-aqueous effect solutions				
Gelatin/air	54.5	Methyl monomer/air	37.0	Methyl monomer/gelatin	4.5	13.0
		Heptyl monomer/air	28.3	Heptyl monomer/gelatin	3.4	22.8

[a]Reprinted with permission of *The Journal of Applied Polymer Science*.

The interfacial tensions obtained in the blood effect solutions as well as in the case of the gelatin solution are rather close in value for both monomers. This indicates that both the methyl monomer, which is more hydrophilic, and the heptyl monomer, which is more hydrophobic, appear to be interacting equally with the protein molecules at the surface of the protein solutions. Although heptyl monomer spreads on both water and blood, it polymerizes slowly on water and instantaneously on blood. This result may be due to two possible factors: (1) the possibly increased concentration of initiating nucleophiles present in blood compared to water (perhaps lone pairs of electrons on the pendant amino groups in proteins may serve as nucleophiles); and/or (2) the protective colloid action of the protein molecules at the blood-monomer interface which may have a tendency to solubilize the monomer at the interface, thereby bringing it in closer contact with the initiating nucleophiles. The latter implies that blood presents a more hydrophobic surface to the heptyl monomer than does water. This could arise as a result of the blood proteins at the interface forming coherent but rather amorphous films; the structures are not exactly that of the natural protein but involve partial denaturation through a spreading out of the various hydrophobic side chains in the protein molecules.

The fact that the methyl monomer forms a lens and does not spread on blood is sufficient to account for its slow polymerization. Here the polymerization is limited to the surface of the droplet. The polymer that forms around the surface of the drop is self-protecting, and further polymerization of the monomer depends upon diffusion of the nucleophiles through the protective barrier.

V. Strength of the Adhesive Bond

Because the higher homologs wet and spread on tissue substrates and the lower homologs do not, it was expected that the adhesive strength of tissue bonds which were prepared with the higher homologs would be greater than that of those formed with the lower homologs. By the same token, it was presumed that on nonproteinaceous substrates the opposite effect would be noted. In order to test this hypothesis for proteinaceous substrates, midline incisions 3 cm in length, extending into the subcutaneous tissues, were made in rats and the incision sealed with 2 drops of various alkyl cyanoacrylate monomers as by Leonard *et al.* (1968) described. The adhered wounds were pulled apart on a table model Instron Testing Machine one hour following wound closure as shown in Fig. 2. For measurement of aluminum-to-aluminum bond strength, adhesive joints were prepared using the standard aluminum jigs as the nonproteinaceous substrate.

The mean tissue wound bond strengths, the standard deviations, and standard error of the mean are summarized in Table IV. In agreement with these data, Heiss *et al.* (1968) have recently reported that *n*-butyl 2-cyanoacrylate forms stronger tissue bonds than methyl 2-cyanoacrylate.

Aluminum-to-aluminum bond strengths for the various homologs are summarized in Fig. 3. It may be observed that in contrast to the results obtained for the bond strength on tissue, the methyl monomer gave the highest bond strength; the bond strengths decreased as the length of the alkoxy group increased. In the aluminum specimens test-

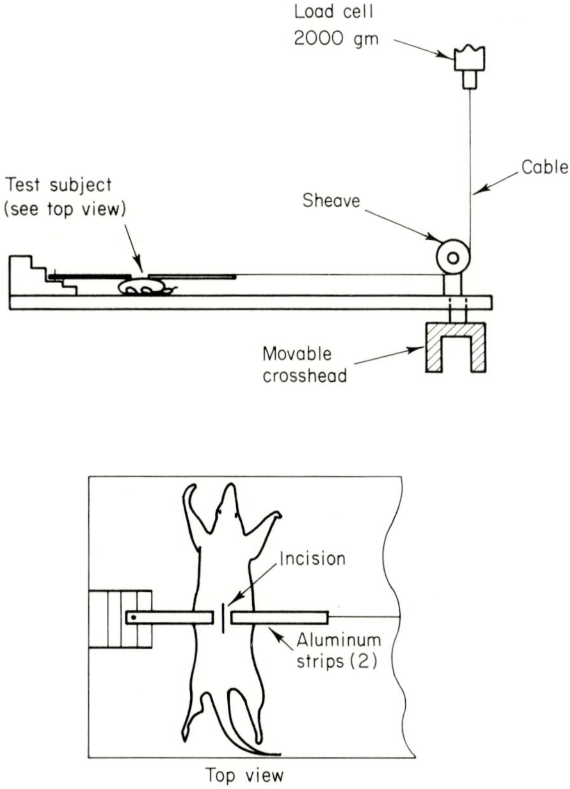

Fig. 2. Testing tissue bond strengths. Reprinted with permission of the *Journal of Biomedical Materials Research.*

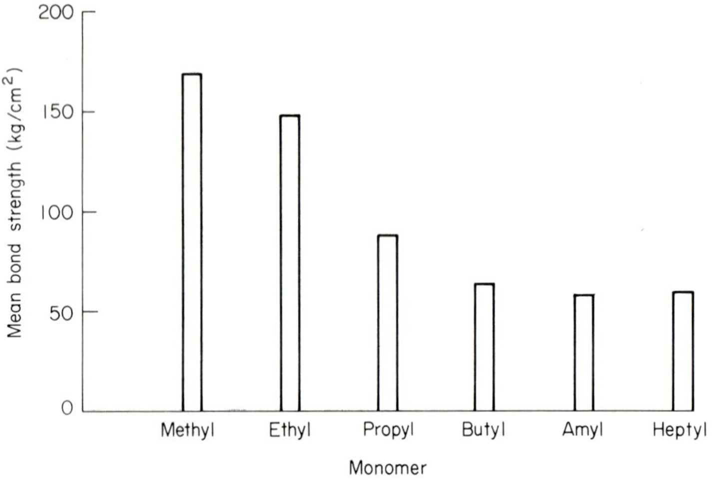

Fig. 3. Aluminum-to-aluminum bond strengths. 24 hour set time. Reprinted with permission of the *Journal of Biomedical Materials Research.*

ed, the strengths obtained were inversely related to the length of the alkoxy chain and considerably less variance in the results were obtained on aluminum substrates than on the tissue substrates.

One possible reason for the general trend toward higher tissue bond strengths as the length of the alkoxy group increases may be the enhanced wetting of the tissue substrates by the higher homologs. That this may not be the only factor responsible for the bond strengths obtained may be inferred from the fact that the standard deviations from the mean bond strengths are quite high. The magnitude of the deviations is not unusual in this type of work when one considers the possible variables in such experiments. These are variations between animals, the variation in bond surface area, the possibility of cohesive failure of the adhesive as well as tissue substrate failure. All these factors may contribute to the variance in the results obtained (see Table IV).

VI. Polymer Degradation

It has been shown in studies with β-^{14}C-tagged methyl α-cyanoacrylate and β-^{14}C-tagged methyl α-cyanoacrylate that the polymers formed *in vivo* are biodegradable. In one instance rats were used as the experimental animal, and in the other, guinea pigs (Cameron, 1965; Reynolds *et al.*, 1966). The data indicated that the metabolities produced from the degradation were eliminated through the normal excretory routes and none of the radioactive degradation products were ultimately stored in the tissues. In implantation experiments in guinea pigs, the tagged polymer had disappeared after 107 days. In rats, the radioactive polymer had substantially disappeared within 154 days. In another experiment, comparison of the rates of disappearance of several 2-cyanoacrylates were made. The data indicate that the methyl homolog degrades at the fastest rate, the *n*-butyl at the slowest, and the polymers with branched alkoxy groups at some intermediate values (see Fig. 4).

The mechanism of the polymer degradation became of interest to determine. It has been shown that the cyanoacrylate polymer degrades in the presence of distilled water with the formation of formaldehyde (Leonard *et al.*, 1966). The process of degradation reaches an equilibrium state at which the amount of formaldehyde produced remains constant. This equilibrium value is reached slowly at pH 7 at 25°C; in neutral boiling water or in cold alkaline dispersion, the equilibrium can be reached more quickly. The results of experiments in which polymer powders were boiled with water at pH 7 are summarized in Fig. 5. The data indicate that the rate of aqueous degradation is considerably slower for the polymers of the higher alkylesters. Polymethyl α-cyanoacrylate degrades much faster than others, and the degradation rate lessens for higher members of the series. It is interesting that rates of degradation *in vitro* are quite analogous to the *in vivo* results.

The study of kinetics of hydrolytic degradation of the polymers in homogeneous solution in acetonitrile in excess water *in vitro* showed that the reaction obeys first order kinetics, indicating a pseudounimolecular reaction (Leonard *et al.*, 1966). The rates of degradation in the

TABLE IV

TISSUE WOUND BOND STRENGTH TESTS[a]

Adhesive	Mean wound strength (gm)	Standard deviation	Standard error	Range	*n*
Methyl	167	55	14	70-315	15
Ethyl	300	127	32	110-530	16
Propyl	298	142	35	140-645	16
Butyl	216	83	20	150-490	17
Amyl	249	119	27	130-530	20
Heptyl	336	171	46	170-615	14

[a]Reprinted with permission of *J. Biomed. Mat. Research.*

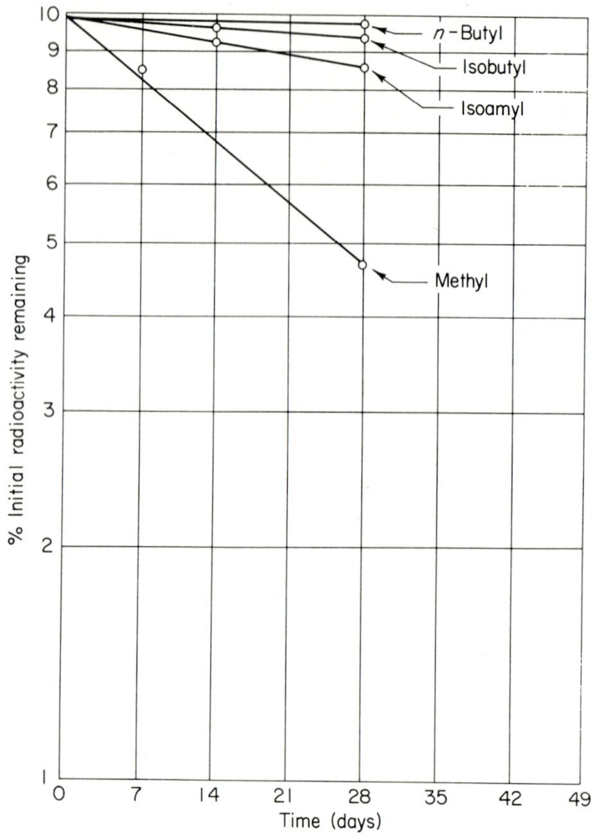

Fig. 4. *In vivo* degradation rates. Reprinted with permission of the *Journal of Biomedical Materials Research.*

homogeneous phase are also much larger for the methyl polymer and are diminished in case of the higher esters at pH 7. The rates of degradation, however, remain high in alkaline solutions (pH 8) for all polymers and differ only slightly from each other (Table V). The data suggest that the degradation mechanism involves an initial attack by hydroxyl ion leading to the reverse Knoevenagel reaction, resulting in the formation of formaldehyde. It is also possible that ester hydrolysis could occur, leading to the formation of water soluble polycyanoacrylic acids which would diffuse away from the implant site.

In vivo studies lend credence to the mechanism of chain scission by hydrolysis. If β-^{14}C-tagged α-cyanoacrylate is subcutaneously im-

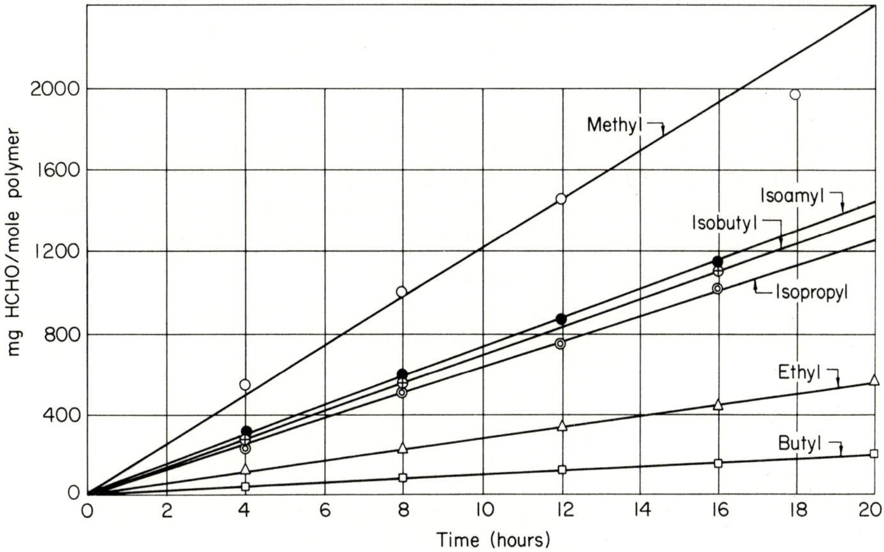

Fig. 5. *In vitro* degradation rates.

planted in rats, radioactive urea is recoverable from the urine. Radioactive urea is not recoverable when α-carbon-tagged cyanoacrylates or the carbon in the nitrile group is tagged (Scullin, 1967). The implication from these data is that in consonance with the *in vitro* findings, radioactive formaldehyde may be formed, and subsequently oxidized to radioactive carbon dioxide which can combine with ammonia to form urea. Another interesting finding is that the *in vitro* rates of degradation in accordance with the proposed mechanism are in the same order as the *in vivo* degradation rates.

TABLE V

RATES OF HOMOGENEOUS DEGRADATION OF CYANOACRYLATE POLYMERS[a]

Polymer of	Number-average molecular weight	K (hr^{-1})	
		pH 7	pH 8
Methyl ester	2204	3.0×10^{-3}	1.0×10^{-2}
Ethyl ester	1533	2.0×10^{-4}	1.5×10^{-2}
Butyl ester	2054	1.0×10^{-5}	2.0×10^{-2}

[a]Reprinted with permission of *The Journal of Applied Polymer Science.*

VII. Mechanism of Hemostasis

The mechanism by which α-cyanoacrylates achieve hemostasis is a matter of conjecture at this time. Because of the formation of a tightly adherent film on the traumatized tissue substrate, it is possible that the cyanoacrylate film causes mechanical blockage to slow blood flow and serves as a foreign surface to activate the clotting cascade. There is preliminary evidence as well that cyanoacrylates may form a porous film which becomes invaded with blood. Subsequent clotting occurs within the film pores. The latter could mechanically enhance the strength of the cyanoacrylate tissue bond (Wade, 1968).

VIII. Medical Implications

Medical evaluation of methyl 2-cyanoacrylate monomer as a tissue adhesive has indicated that the polymer as well as the monomer elicits acute inflammatory responses in tissues. Further medical studies with the homologous series of α-cyanoacrylates prepared in this laboratory have indicated that as the homologous series is ascended, the inflammatory response is decreased. The butyl derivative and higher homologs appear to be well tolerated by the tissues. These data are in general accord with the *in vitro* studies presented in this chapter. If indeed the implanted polymers degrade *in vivo* as has been demonstrated *in vitro*, then, in the vicinity of the implanted polymer particles, one may expect to find formaldehyde and an alkylcyanoacetate (if the degradation proceeds to the ultimate stoichiometric products). Both these compounds are toxic and can elicit acute inflammatory responses. The fact that the butyl derivative is tissue-tolerated and degrades at a considerably slower rate than the methyl derivative, which is not well tolerated by the tissues, implies that the tissues can more easily metabolize the lower concentration of degradation products present at a given time in the case of the butyl derivative.

The implications of these findings for clinical application of the cyanoacrylates are readily apparent. It seems that the longer side-chain members not only exhibit the least acute toxicity, but also are the most effective in achieving hemostasis, easiest to apply, and form stronger tissue adhesive bonds. One factor in favor of the methyl α-cyanoacrylate, however, is its more rapid disappearance from the body. The salutary combination of effectiveness in hemostasis-inducing ability of the higher homologs and the rapid biodegradation of the methyl monomer would be highly desirable in tissue adhesives.

References

Cameron, J. L. (1965). *Surgery* **58**, 2.

Coover, H. W., Jr., Joyner, F. B., Shearer, N. H., and Wicke, T. H. (1959). *SPE (Soc. Plastics Engrs.) J.* **15**, 413.

Heiss, W., Guthy E., and Faul, P. (1968). *Programme Abstr. Symp. Adhesives Surg., 1967* p. 84.

Jeremias, C. G. (1956). U.S. Patent 2,763,677 (to Eastman Kodak).

Kulkarni, R. K. (1967). Private communication.

Leonard, F., *et al.* (1966). *J. Appl. Polymer Sci.* **10**, 259-272.

Leonard, F., Hodge, J. W., Jr., Houston, S., and Ousterhout, D. K. (1968). *J. Biomed. Mater. Res.* **2**, 173-178.

Reynolds, R. C., Fassett, D. W., Astill, B. D., and Casarett, L. J. (1966). *J. Surg. Res.* **6**, 5.

Scullin, R. (1967). Private communication.

Wade, C. W. R. (1968). Private communication.

CHAPTER 12

BUTYL CYANOACRYLATE AS A SURFACE ADHESIVE IN HUMAN ORAL WOUNDS

Surindar N. Bhaskar

Many members of a family of chemical adhesives with a general formula of CH= C (CN) − COOR have the ability to adhere to and cement together moist living tissues (Coover *et al.*, 1959). It has been shown that these substances are hemostatic agents and that, when embedded in deep tissues, they are partly phagocytosed by histiocytes (Bhaskar *et al.*, 1966a,b, 1967a; Bhaskar and Frisch, 1968; Bhaskar, 1968a). When applied on the surface of the mucous membrane or skin, however, they are spontaneously exfoliated (Bhaskar and Frisch, 1968; Bhaskar and Cutright, 1969). In addition to their hemostatic and adhesive properties they have been shown to be bacteriostatic (Bhaskar *et al.*, 1966b; Bhaskar and Frisch, 1968; Lehman *et al.*, 1966). A comparative analysis of a number of chemical adhesives of the cyanoacrylate family showed that the normal butyl was biologically best tolerated (Bhaskar *et al.*, 1966a, 1967a). Since the cyanoacrylates when applied as a surface dressing were found to exfoliate and to cause no undesirable tissue reaction in the rat and the monkey (Bhaskar, 1968b), they were tested in a variety of oral surgical procedures in man. The immediate adhesion of this material to the *cut* tissues and the concomitant hemostasis (probably by sealing the capillary walls) were its most striking properties. It was observed during its application in man and animals that it did not adhere to epitheliated oral mucosa and this made it easier to remove from the nonsurgical areas. On cut surfaces, butyl cyanoacrylate polymerized immediately and adhered to the area for varying lengths of time. Since this material is somewhat brittle, its retention to cut tissues is determined by their mobility. In highly mobile areas such as the floor of the mouth it becomes dislodged rapidly while on the gingiva it lasts up to 7 or more days.

In dental practice butyl [and isobutyl (Bhaskar, 1968a)] cyanoacrylate have been used under several clinical conditions.

I. Periodontal Surgery

The response of surgically treated gingiva to the spray of butyl cyanoacrylate was consistently excellent (Fig. 1). Following gingivectomy, the butyl cyanoacrylate spray always led to cessation of hemorrhage and the area appeared "frosted" (Fig. 1). On occasion, slow oozing of blood from the interdental area persisted, but this could usually be stopped by spraying the area for the second time. In contrast to the control side where the conventional dressings were used, the cyanoacrylate procedure was much faster and much neater (Bhaskar and Frisch, 1968). Patients generally complained of less postoperative pain, and clinically the operated area had less granulation tissue than the control side. Clinical healing (epithelization) of the gingiva occurred faster than in the area where conventional dressing were used (Bhaskar *et al.*, 1966b). The cyanoacrylate dressing usually lasted 7 days, at which time a second application could be easily made. As the underlying tissues healed the adhesive was either exfoliated or could easily be lifted from the area (Figs. 2-4). In some cases gingivectomy has to be performed around isolated teeth. In such cases it is difficult or impossible to retain the usual dressings. With butyl cyanoacrylate, however, the gingival wounds could be "dressed" with ease, and healing occurred uneventfully.

II. Extraction Sites

The adhesive properties of the butyl cyanoacrylate on tooth extraction sites were studied in rats (Bhaskar *et al.*, 1967b). It was shown that, because of its adhesion to the underlying tissues, it protects the underlying tissues from secondary infection. Although at this time cyanoacrylate has been used on only a limited number of human extraction wounds, initial observations reveal that this material acts as an immediate hemostatic agent and that the wounds heal as fast as, or faster than, when they are left open to the oral cavity.

III. Ulcerations

Butyl cyanoacrylate was also used as an adhesive protective covering on superficial ulcerations of the oral mucosa. These patients had multiple ulcerations of recurrent aphthae or lesions which were secondary to leukemia. Ulcerations ranged in size from 4×4 mm to 50×20 mm. Unlike the wounds of periodontal and oral surgery, they were usually located on the movable portions of the oral mucous membrane (cheeks, vestibule, and lips). Butyl cyanoacrylate did not adhere as well to the mobile oral mucosa as it did to the firm nonmobile areas, and the material was lost on an average of 2 days later. During this period, however, the patients reported relief from pain and were able to eat with less discomfort.

IV. Free Mucosal Grafts

In grafting or transplanting soft tissues in the oral cavity, use of sutures and periodontal packs has been considered indispensable (Nabers, 1966; Bhaskar and Frisch, 1967). Because of the excellent adhesion which the butyl cyanoacrylate provides, this material has been shown not only to be an effective agent in immobilizing the graft at the recipient site but has markedly simplified the technique for this procedure (Bhaskar and Frisch, 1967).

V. Biopsy Sites

Butyl cyanoacrylate has been used to cover the biopsy sites, and in some areas of the mouth, such as the palate, there is no other way by which such wounds can be adequately protected (Bhaskar and Frisch, 1968).

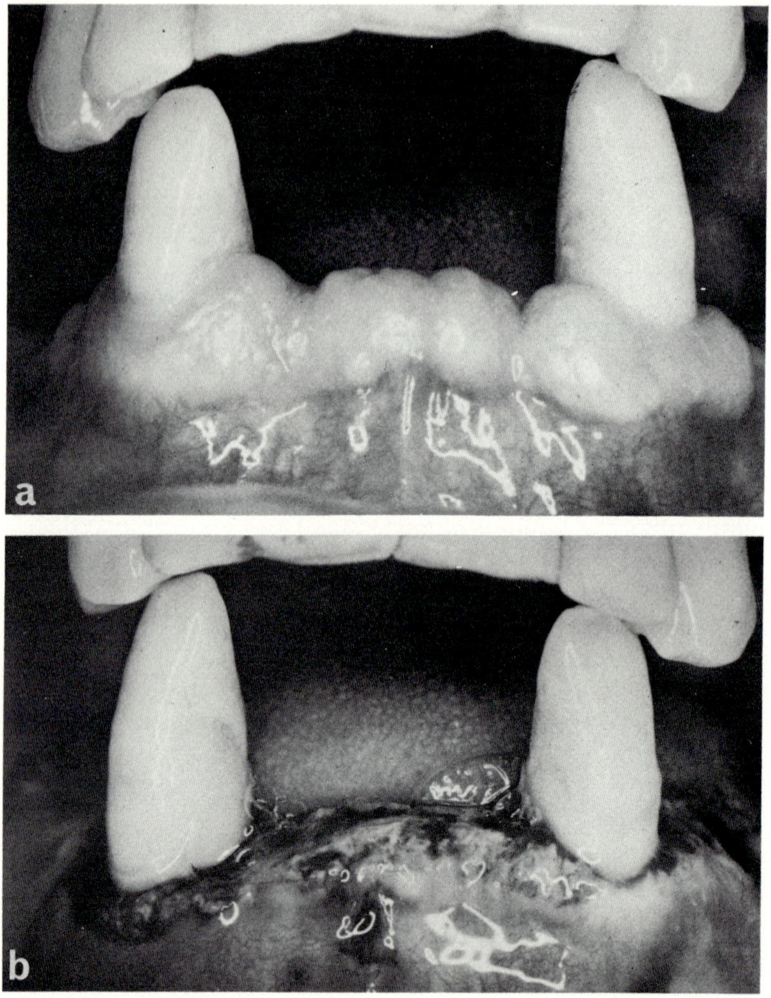

Fig. 1. Four stages in the healing of oral wounds under an application of a tissue adhesive (butyl cyanoacrylate): preoperative (a); after surgery (b); after application of the adhesive (c); and after healing (d).

VI. Sutured Tissue

This material has been used as a protective covering on sutured areas (Bhaskar and Frisch, 1968). In this manner it appears to reduce pain and promote healing.

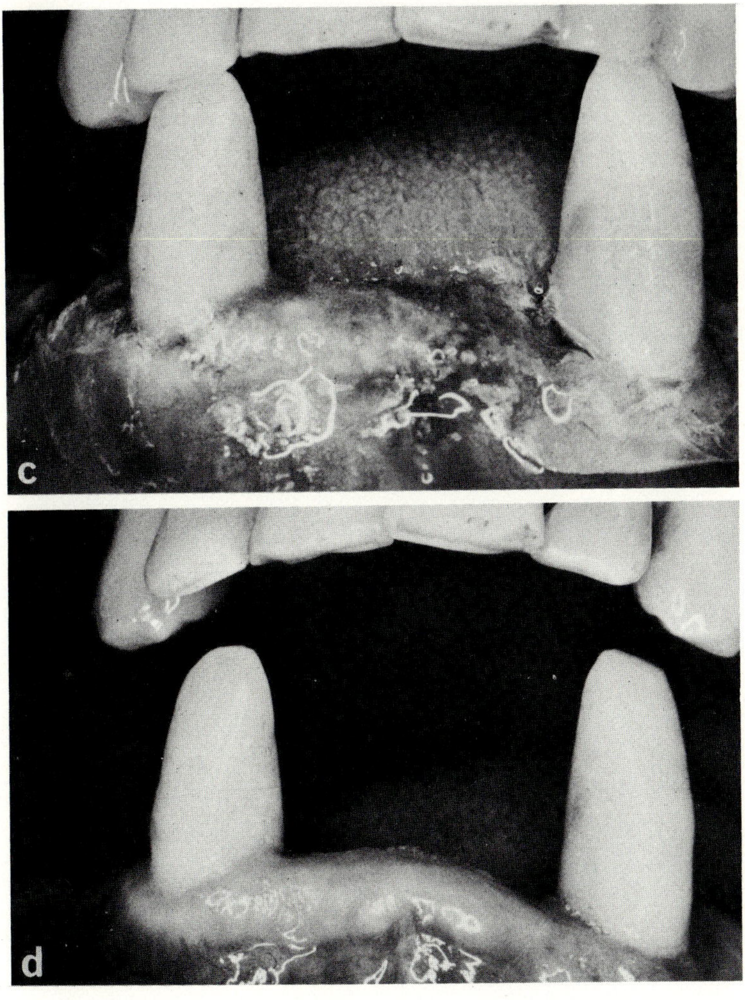

Fig. 1. (c) and (d)

VII. Other Surgical Sites

Cases of inflammatory papillary hyperplasia were treated by curettage which was followed by the cyanoacrylate spray. This method of dressing was found preferable to any in current use.

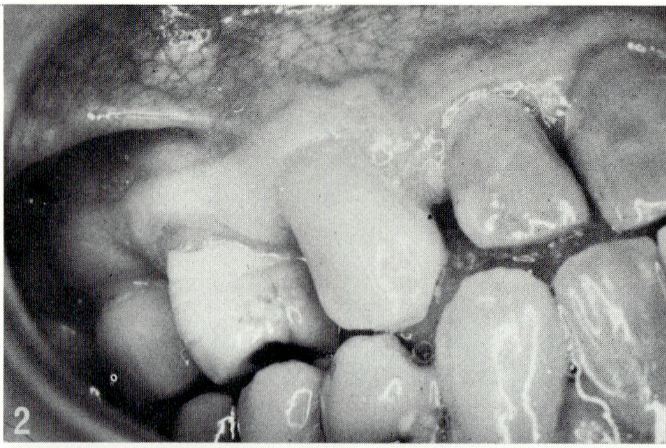

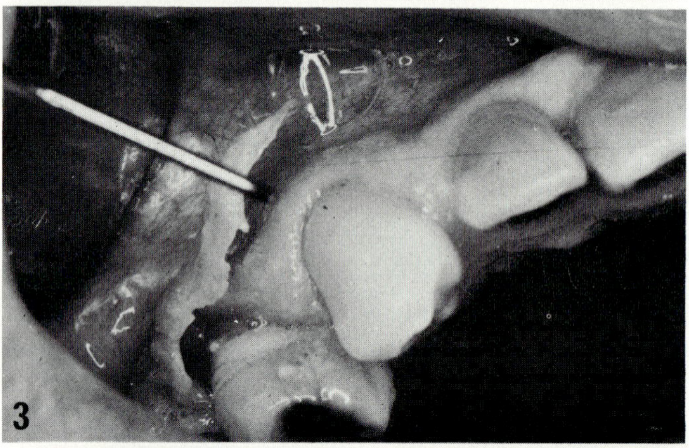

Fig. 2. A case where an adhesive was applied to the gingiva ten days previously.
Fig. 3. How the adhesive can be easily lifted from the healed gingival gingiva.
Fig. 4. The healed gingival gingiva.

VIII. Filling Material for Carious Teeth

When mixed with inert fillers the cyanoacrylates are known to produce hard masses. In this combination butyl cyanoacrylate was tested as a dental restorative material (Bhaskar *et al.*, 1969). Although biologically acceptable to the dental tissues, it washes out in the oral fluids and is not suitable for this purpose (Bhaskar *et al.*, 1969).

IX. Pulp Capping

The adhesive qualities of the butyl cyanoacrylate have been used to protect bleeding or nonbleeding exposed pulps of human and animal teeth (Bhaskar *et al.*, 1970). In this location it preserves the integrity of the dental pulp and protects it from secondary inflammation. The close contact the cyanoacrylate makes with the underlying tissues is illustrated in Fig. 5. It is probably because of this quality that it protects the healing oral tissues from secondary infection.

Fig. 5. Photomicrograph of an interface between viable connective tissue and the adhesive (clear area). Note the absence of any space between the two. ×110 (H&E stain).

X. Summary

The tissue response to the butyl cyanoacrylate, method of application, safety, and clinical uses have been the subject of numerous previous publications (Bhaskar *et al.*, 1966a, 1967a,b, 1969, 1970; Bhaskar and Frisch, 1967, 1968; Bhaskar, 1968a,b; Bhaskar and Cutright, 1969). In these reports it has been shown that (1) the butyl cyanoacrylate and isobutyl cyanoacrylate are well accepted by mammalian tissues; (2) they are hemostatic agents; (3) they act as tissue adhesives; (4) they have bacteriostatic properties and can be used as surface dressings of oral wounds. When used in this manner they reduce pain and secondary infection and promote healing; and (5) they are easily applied.

It appears from these studies that the most important properties which make these substances highly promising in oral surgery are their tissue acceptability and tissue adhesion. It is because of the latter quality that they can act as hemostatic agents as well as are able to prevent the secondary infection of the underlying tissues.

References

Bhaskar, S. N. (1968a). *Oral Surg., Oral Med., Oral Path.* **26**, 573.

Bhaskar, S. N. (1968b). "Use of Normal Butyl Cyanoacrylate in the Oral Cavity; Experimental and Clinical Studies," Ann. Rept. R&D Command, Army Med. Serv., Dept. of the Army.

Bhaskar, S. N., and Cutright, D. E. (1969). *J. Dental Res.* **48**, 294–297.

Bhaskar, S. N., and Frisch, J. (1967). *J. Periodontol.* **39**, 190–195.

Bhaskar, S. N., and Frisch, J. (1968). *J. Am. Dental Assoc.* **77**, 831–837.

Bhaskar, S. N., Jacoway, J. R., Margetis, P. M., Leonard, F., and Pani, K. C. (1966a). *Oral Surg., Oral Med., Oral Pathol.* **22**, 394–404.

Bhaskar, S. N., Frisch, J., Margetis, P. M., and Leonard, F. (1966b). *Oral Surg., Oral Med., Oral Pathol.* **22**, 526–535.

Bhaskar, S. N., Frisch, J., Margetis, P. M., and Leonard, F. (1967a). *Oral Surg., Oral Med., Oral Pathol.* **24**, 137–144.

Bhaskar, S. N., Frisch, J., and Cutright, D. E. (1967b). *Oral Surg., Oral Med., Oral Pathol.* **24**, 604–616.

Bhaskar, S. N., Frisch, J., and Margetis, P. M. (1969). *J. Dental Res.* **48**, 57–60.

Bhaskar, S. N., Cutright, D. E., Boyers, R. C., and Margetis, P. M. (1970). *J. Am. Dental Assoc.*

Coover, H. W., Jr., Joyner, F. B., Shearer, N. H., Jr., and Wicker, T. H., Jr. (1959). *SPE Soc. Plastics Engrs. J.* **15**, 413.

Lehman, R. A., West, R. L., and Leonard, F. (1966). *Arch. Surg.* **93**, 447.

Nabers, J. M. (1966). *Periodontics* **4**, 243.

CHAPTER 13

REVIEW OF CYANOACRYLATE MONOMERS IN SURGERY*

Teruo Matsumoto

Large wounds of solid organs such as liver and kidney occasionally present serious surgical problems and may even result in death from uncontrollable hemorrhage. At present, the usual procedure recommended and employed for treating these wounds is the use of suture ligatures with or without gelfoam or other hemostatic agents and of adequate drainage. These methods are not always successful or satisfactory, and often are time consuming. Because of the limitations in these techniques, various surgical repairs of injured internal organs using aerosol tissue adhesive spray were attempted.

Various cyanoacrylate monomers and a formaldehyde-gelatin-resorcinol tissue adhesive (Matsumoto *et al.*, 1968c) were evaluated in 400 dogs, 200 rats, and 200 white mice. At present, it appears that the best tissue adhesives are *n*-butyl and isobutyl cyanoacrylate monomers.

*The principles of laboratory animal care as promulgated by the National Society for Medical Research were observed.

The spray gun (Leonard *et al.*, 1965) used in these studies permitted delivery of a fine spray which formed a homogeneous, uniform surface film with a minimum volume of adhesive. However, using the spray gun required caution, a source of compressed gas was required to apply the spray, and cleaning the nozzle and Teflon tip with nitromethane after each procedure was time consuming. Sterilization of the spray gun and equipment is not simple. In addition to these disadvantages, the gun occasionally malfunctions at the time when the spray is needed.

A sterile, individually disposable aerosol tissue adhesive spray (Matsumoto *et al.*, 1967a,b) has been developed at the Division of Surgery, Walter Reed Army Institute of Research, with the assistance of G. Barr Company, Niles, Illinois. Success in surgical application of the aerosol tissue adhesive spray has depended on the development of a more suitable monomer and surgical technique (Matsumoto *et al.*, 1967c,f, 1968a,b). Present surgical techniques using suture materials must be modified when aerosol tissue adhesive spray is used.

This chapter discusses the techniques of hemostasis and repair of hepatic and renal wounds, reinforcement of vascular anastomosis, intestinal anastomosis, closure of intestinal perforation, and the sealing of resected lung with aerosol tissue adhesive spray.

I. Experimental Spray

A. *Hemostasis of Liver Wound (Matsumoto, 1967c; Matsumoto et al., 1967g)*

During laparotomy of heparinized dogs, a 5 × 5 cm wound of a lobe of the liver was created by excising tissue. The vascular supply to the liver was temporarily occluded by applying a vascular clamp across the portal triad, and the wound was dried by sponging. A thin polyethylene sheet with a window was applied around the wound to protect the remaining surgical field from contact with the tissue adhesive. Aerosol tissue adhesive was sprayed on the liver wound, and hemostatis was confirmed after removing the vascular clamp. Surgery was completed by removing the thin polyethylene sheet and closing the abdomen.

B. *Repair of Injured Kidney (Matsumoto et al., 1967h, 1968f)*

Following anticoagulation of the animal with heparin, the left kidney was mobilized and a vascular clamp was applied to the renal vessels. A thin, windowed, polyethylene sheet was applied around the kidney and used to protect the surrounding area and organs. The renal cortex and medulla were incised along Brodel's line deep enough to expose the calyces thoroughly. Experimental studies at the Walter Reed Army Institute of Research indicated that the inadvertent introduction of cyanoacrylate monomer into renal calyces predisposed to calculus formation.

A second, smaller, thin polyethylene sheet was used to cover the calyces. Accumulated blood between the cut surfaces of the kidney was sucked out and the oozing blood was sponged gently with gauze. The cut surfaces had to be reasonably dry so that binding would occur. Aerosol tissue adhesive spray was sprayed on the cut surfaces and the two surfaces were brought in contact with gentle manual pressure. The kidney halves were held together for 10–20 seconds, during which time the polymerization of the monomer occurred. The vascular clamp was removed and complete hemostasis was confirmed. The polyethylene sheet was removed and the kidney was replaced in its original position.

C. Sealing of Resected Lung (Matsumoto et al., 1967i)

A right thoracotomy was performed at the fourth intercostal space, and the apical lobe was indentified. A noncrushing clamp was placed across the lobe, and a wedge resection was performed. A thin polyethylene sheet was placed on the entire surgical field except for the cut surface of the lung. The cut surface of the lung was sponged and dried. Aerosol tissue adhesive was sprayed on the surface. Gentle and constant pressure was applied to this area through a layer of polyethylene sheet to avoid adhesion of the tissue adhesive to the glove. Adequate time was allowed for polymerization. The noncrushing clamp and polyethylene sheet were removed, and the lung was inflated to test for leakage of air and blood.

D. Intestinal Anastomosis (Matsumoto et al., 1967e, 1969b)

Our 16-month follow-up study indicated that anastomosis of tubular organs was more rapid at first but resulted in increased mortality and complication rate compared to the conventional suture (Matsumoto *et al.*, 1967j). At present, it is apparent that the best use of aerosol spray in surgery of tubular organs is for reinforcement of suture lines. Conventional suture lines were reinforced by aerosol tissue adhesive after protection of the remaining surgical field. Isotonic saline was injected into the closed loop between the clamps to prove security of the anastomosis.

E. Closure of Perforated Intestine (Matsumoto et al., 1967k)

In the 16-month follow-up, primary patch closure of perforations with tissue adhesive alone revealed increased morbidity compared to the conventional suture. Aerosol patch closure was performed with conventional suture. The surgical field was covered with a polyethylene sheet except for the suture lines, which were sprayed with aerosol tissue adhesive.

A perforated segment of intestine was compressed firmly against the normal segment of intestine. Isotonic saline was injected in order to test the security of the patch closure.

F. Vascular Anastomosis (Matsumoto et al., 1967j, 1968d)

Following anticoagulation, the femoral artery was divided between two vascular clamps. Conventional suture anastomosis was completed. However, leakage of blood from the suture lines was extensive. The artery on both sides of the suture line was then clamped by vascular clamps and the surgical field was protected with a thin polyethylene sheet. *n*-Butyl or isobutyl cyanoacrylate monomer was dropped on the suture lines. Complete hemostasis and functional anastomosis were noted immediately. Intimal deposition of all available cyanoacrylate monomers caused thrombosis, and the potency rate for primary suture anastomosis was greater than that for primary cyanoacrylate anastomosis.

G. Follow-up Study

No untoward side effects or complications were noted when aerosol tissue adhesive spray was used for hemostasis of injured solid organs or reinforcement of leaking anastomosis. Minimal tissue reaction and residual polymer fragments were seen in all cases treated with *n*-butyl or isobutyl monomers and sacrificed 24 hours to 16 months after surgery (Matsumoto *et al.*, 1967d).

Lamborn and others (1970) at the Division of Surgery, Walter Reed Army Institute of Research have shown that the tensile strength of skin bound with various cyanoacrylate polymers is greater than that of sutured skin within the first 72 hours; this is apparently due to the tensile strength of the polymer itself. Vital microscopic and microangiographic studies suggested that the increased tensile strength which accompanies the natural wound healing related to the availability of maximal surface area, across which proliferating fibroblasts can effect a cellular and eventually collagenous union (Matsumoto *et al.*, 1969c).

Because of the long biological half-life of the *n*-butyl and the longer chain cyanoacrylate polymers, it is important to study whether they may be collagenic or have other untoward effects (Matsumoto, 1967a,b). In 400 dogs with *n*-butyl and isobutyl cyanoacrylate monomer applications to wounded internal organs, no tumors formed over an observation of 22 months; nor were any tumors found in the 200 rats and 200 white mice followed for the same period of time (Matsumoto *et al.*, 1969a).

The freon propellants introduced by aerosol spray were found to remain in the body for a short period. This resulted from physical trapping of Freon in the polymer film and not from any chemical change in the monomer. Freon trapped in the polymer appeared to be absorbed by the blood and eliminated from the lung (Matsumoto *et al.*, 1968e).

II. Clinical Application

Since the higher homologous monomers disappear from the tissue very slowly and long-term study of the liquid propellant used in the internal organs is not completed, a longer period of observation is necessary to determine the presence of untoward side effects, particularly carcinogenicity and/or abscess formation. At present, the clinical use of aerosol tissue adhesive spray should be restricted. Indications for their use appear to be in control of hemorrhage, especially from the solid organs as occurs with trauma, surgery of liver and kidney, and in patients with bleeding diathesis. Other uses need further evaluation. Tissue adhesive has been used successfully for hemostasis for otherwise fatal hemorrhage in 29 patients with mortal injuries to the liver, kidney, major arteries, or retroperitoneal spare. Seven members of the U.S. Armed Forces in Vietnam have been followed from 5 to 19 months without evidence of untoward sequela. A detailed long-term study of these patients is essential.

III. Summary

Surgical techniques have been developed for hemostasis of solid organs and reinforcement of leaking suture line using aerosol tissue adhesive spray or drop method. Drying of the area by temporary clamping of blood supply to the organ and protection of the surrounding surgical field with a thin polyethylene sheet were two essential steps before application of the spray.

Although preliminary studies suggest that these substances are noncollagenic or teratogenic for laboratory animals, their long-term effect on human tissues is not known. Therefore, *n*-butyl and isobutyl cyanoacrylate monomers in Freon spray or drop method should be used clinically only in an emergency situation as a lifesaving procedure when all conventional surgical methods have failed.

References

Lamborn, B. Th., Soloway, H. B., Matsumoto, T., and Aaby, G. V. (1970). Unpublished data.

Leonard, F., Bushey, T. J., and Cameron, J. L. (1965). *Surgery* **57**, 749.

Matsumoto, T. (1967a). *J. Am. Med. Assoc.* **201**, 195.

Matsumoto, T. (1967b). *J. Am. Med. Assoc.* **202**, 145.

Matsumoto, T. (1967c). *Military Med.* **132**, 951.

Matsumoto, T., Hardaway, R. M., Pani, K. C., and Margetis, P. M. (1967a). *Arch. Surg.* **95**, 685.

Matsumoto, T., Hardaway, R. M., Heisterkamp, C. A., Pani, K. C., Leonard, F., and Margetis, P. M. (1967b). *Am. Surgeon* **33**, 671.

Matsumoto, T., Hardaway, R. M., Heisterkamp, C. A., Pani, K. C., Leonard, F. M., and Margetis, P. M. (1967c). *Arch. Surg.* **94**, 858.

Matsumoto, T., Hardaway, R. M., Heisterkamp, C. A., Pani, K. C., and Leonard, F. (1967d). *Arch. Surg.* **94**, 861.

Matsumoto, T., Hardaway, R. M., Pani, K. C., Leonard, F., Heisterkamp, C. A., and Margetis, P. M. (1967e). *Surgery* **61**, 567.

Matsumoto, T., Pani, K. C., Hardaway, R. M., and Leonard, F. (1967f). *Arch. Surg.* **94**, 153.

Matsumoto, T., Pani, K. C., Hardaway, R. M., Leonard, F., and Heisterkamp, C. A. (1967g). *Arch. Surg.* **94**, 187.

Matsumoto, T., Pani, K. C., Hardaway, R. M., Leonard, F., Jennings, P. B., and Heisterkamp, C. A. (1967i). *Surgery* **61**, 573.

Matsumoto, T., Pani, K. C., Hardaway, R. M., Leonard, F., Jennings, P. B., and Teschan, P. E. (1967j). *Arch. Surg.* **94**, 388.

Matsumoto, T., Pani, K. C., Hardaway, R. M., and Leonard, F. (1967k). *Arch. Surg.* **94**, 184.

Matsumoto, T., Dubek, A. S., Pani, K. C., Kovaric, J. J., and Hamit, H. F. (1968a). *Arch. Surg.* **97**, 727.

Matsumoto, T., Hardaway, R. M., and Pani, K. C. (1968b). *Am. Surgeon* **34**, 263.

Matsumoto, T., Pani, K. C., and Hamit, H. F. (1968c). *Military Med.* **134**, 98.

Matsumoto, T., Pani, K. C., and Hamit, H. F. (1968d). *Arch. Surg.* **96**, 405.

Matsumoto, T., Pani, K. C., Kovaric, J. J., and Hamit, H. F. (1968e). *Arch. Surg.* **97**, 727.

Matsumoto, T., Pani, K. C., Margetis, P. M., and Hamit, H. F. (1968f). *Urol. Dig.* **7**, 27.

Matsumoto, T., Nemhauser, G. M., Soloway, H. B., Heisterkamp, C. A., and Aaby, G. V. (1969a). *Military Med.* **134**, 247.

Matsumoto, T., Pani, K. C., Kovaric, J. J., and Hamit, H. F. (1969b). *Am. Surgeon* **35**, 210.

Matsumoto, T., Soloway, H. B., Cutright, D. E., and Hamit, H. F. (1969c). *Arch. Surg.* **98**, 266.

CHAPTER 14

ADHESION TO SKIN

Effects of Surgical Tapes on Skin

James W. Bothwell

I. Introduction

During the past decade major advances have been made in the performance of surgical tapes. Until 1962 the standard for surgical tapes had been rubber based adhesive on a cotton cloth backing. In recent years acrylic adhesives on rayon cloth or on a nonwoven fabric backing have become popular. The dominant reason for the broad acceptance of acrylic adhesives has been the so called "hypoallergenic" nature of the newer tapes, which has eliminated the large number of components of the rubber tapes: crepe or synthetic rubber, natural rosins, antioxidants, plasticizers, filling and coloring agents. The fact that many of these materials come from plant sources leads to an inherent problem of source variation and to a potential for introduction of irritant and sensitizing agents.

However, many of the problems attributed to "sensitizing" are not necessarily related to the components of the rubber adhesive, but rather to the changes in cohesion of horny layer cells produced by wearing the tape. Apparently an environmental alteration allows overhydration and softening of the outer layers of skin to occur. When the tape is removed, variable planes of separation within the skin develop, and the separation interface occurs not at the skin-adhesive interface but deeper in the stratum corneum.

II. Experimental Tape Stripping of the Skin

Pressure sensitive adhesive tapes have been used by biologists and dermatologists for many years as a tool to explore the character of the cornified layers of the epidermis. Tape stripping of the skin was described originally by Wolf (1939) and popularized by Pinkus (1951) in the United States. Goldschmidt and Kligman (1967) have used a modified stripping procedure in which the adhesive is applied directly to a slide and microscopically examined to characterize the cells of the stratum corneum. They have also explored the removal of horny cells per sequential strip of adhesive tape. Kligman concluded that since fewer cells per square millimeter were removed with each tape strip, there is an increasing cellular cohesion gradient from the surface downward. Tregear and Dirnhuber (1962) estimated the quantity of cells on tapes using a dye uptake technique and have come to similar conclusions. Results with stripping of rat skin were similarly reported by Jenkins and Tresise (1969). Kligman (1964) noted that the wet tensile strength of the horny layer is only a fraction of that in the dry state.

Figure 1 demonstrates the effect of serial stripping on an area of back skin overhydrated for 24 hours by Saran occlusion, and on an untreated control area. In contrast to the total of 6.8 μg (dry wt.) of stratum corneum removed with six sequential strips over the control area, 17.2 μg (dry wt.) or more than twice the amount is removed from the overhydrated area. This effect is especially noticeable in the first few strips and is similar to the situation encountered with the use of an occlusive surgical tape.

When an occlusive tape is left in place for an extended period of time (Fig. 2), the force necessary to remove it from the skin rapidly builds up within 4 hours and then erratically falls to a much lower level. Inspection of removed tapes subjected to the dye uptake procedure (Tregear and Dirnhuber, 1962) shows that large quantities of stratum corneum adhere to the adhesive, indicating separation deep in the cornified layers. In contrast, more porous tape shows a somewhat slower buildup of adhesion and smaller amount of stratum corneum on the adhesive. This indicates that the plane of separation is higher in the stratum corneum and corroborates the observation reported by Orentreich *et al.* (1966) for occlusive and nonocclusive tapes.

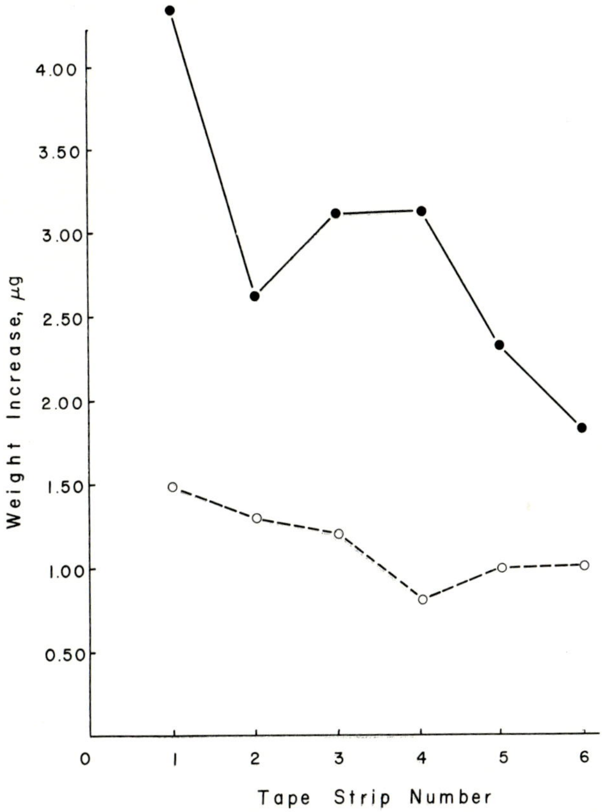

Fig. 1. Comparison of weight of horny cells removed from back skin overhydrated by 24 hour Saran occlusion and from adjacent untreated control skin. Identical segments of predried pressure sensitive tape were individually weighed on a microbalance, evenly placed over the hydrated or the control area, left in place for 1 minute, removed, dried, and reweighed. Total weight from hydrated area (●——●) was 17.2 μg and from untreated control area (O---O), 6.8 μg for six strips.

III. Measurements of Force of Removal

Experiments in which force of removal is measured for serial application of tape to a given skin area offers another perspective to the effect of overhydration on the stripping of skin. When normal skin is sequentially stripped to the glistening Malpighian layer (Fig. 3), the force of removal of each tape increases as lower and lower cell layers in the stratum corneum are reached. In this region the cohesion of the epidermal cells is greater than the adhesion to the stripping tape.

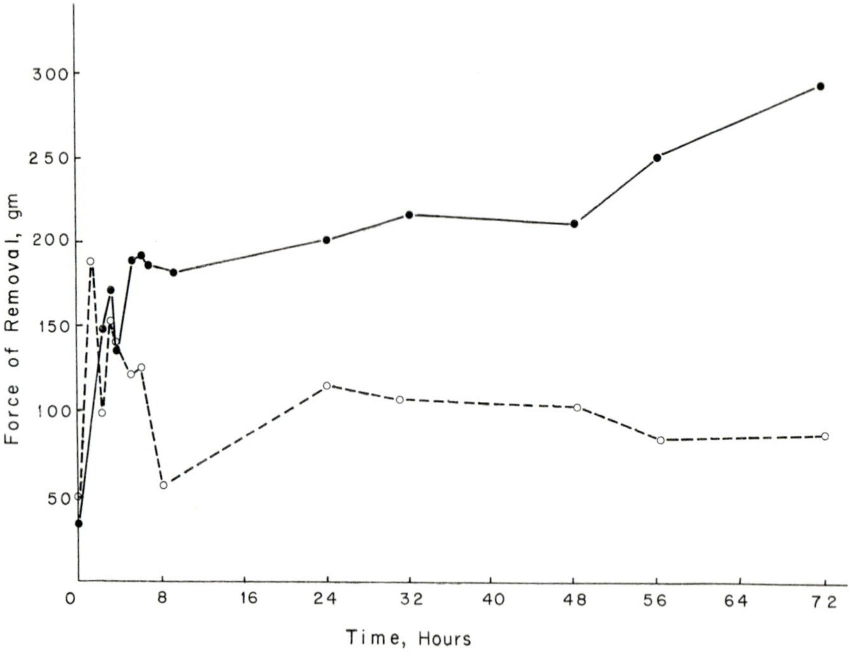

Fig. 2. Force of removal of 1 inch × 2 inch segments of a nonocclusive (●——●) tape (acrylic adhesive-rayon taffeta backing) and an occlusive tape (O---O) (nonperforated rubber adhesive-cotton cloth backing) from back of volunteer subject. Tapes were worn for periods up to 3 days.

The initial rapid buildup of adhesion of occlusive tape and subsequent lower removal forces is a typical example from a much larger group of subjects. The somewhat slower buildup of adhesion and continued high level for nonocclusive tape is also typical.

Beyond this plateau region the force diminishes to a steady state value where the glistening layer has been reached and cells are no longer being removed.

The rate of increase of adhesion from strip to strip in such an experiment is highly variable from subject to subject (Table I), as is the strip number at which the maximum removal force is reached before diminishing. The shape of the curve, however, is consistent. Jenkins and Tresise (1969) as well as Kligman (1964) described some of the reasons for unevenness of tape stripping.

A similar picture of increased adhesion as the stratum corneum is removed and lower cell layers are reached is seen (Fig. 4) when an occlusive tape is worn for 24 hours, removed, a fresh sample replaced over the same site, and the sequence repeated daily for 7 days. Examination of the adhesive face of the dyed tapes revealed that large amounts of stratum corneum adhered to the occlusive adhesive, particularly at days 1–3.

The more porous tape, however, reaches maximum adhesion within the first 24 hours. Daily application does not seem to change the force of removal for at least 7 days. Only small amounts of stratum corneum

TABLE I

Increase of Adhesion with Sequential Stripping of Human Back Skin

Subject	Average rate of force increase per strip (gm)	Strip no. at which maximum adhesion reached
L.D.	12.4	13
J.P.	12.0	19
J.S.	7.4	18
J.B.	13.6	16
R.K.	9.8	21
R.P.	10.6	15

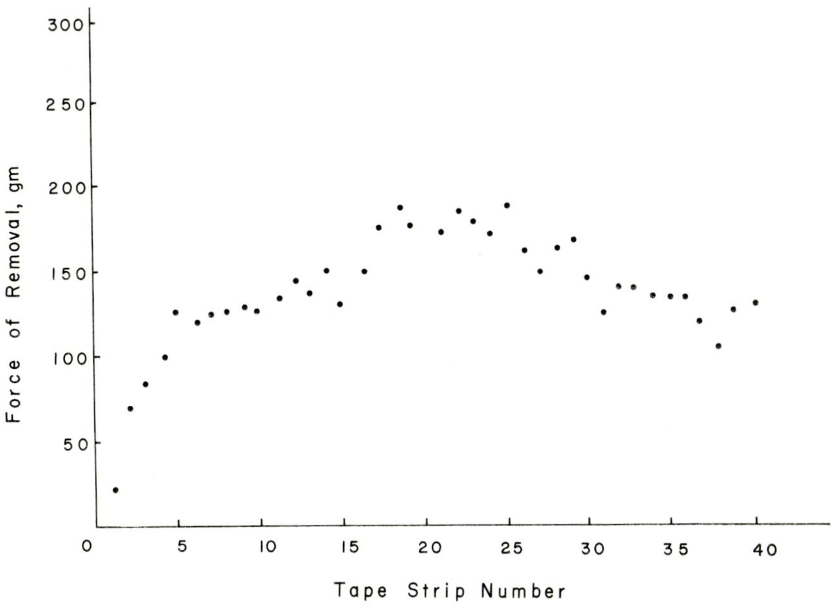

Fig. 3. Force of removal of serial strips of 1 inch × 2 inch segments of pressure sensitive tapes left in place for 1 minute on a defined area of human back skin. Tapes were removed at a rate of 12 inches per minute with hydraulic drive using an LVDT-strain gauge unit with chart readout. Force is the mean removal force expressed in grams. The glistening layer was reached at about strip no. 30.

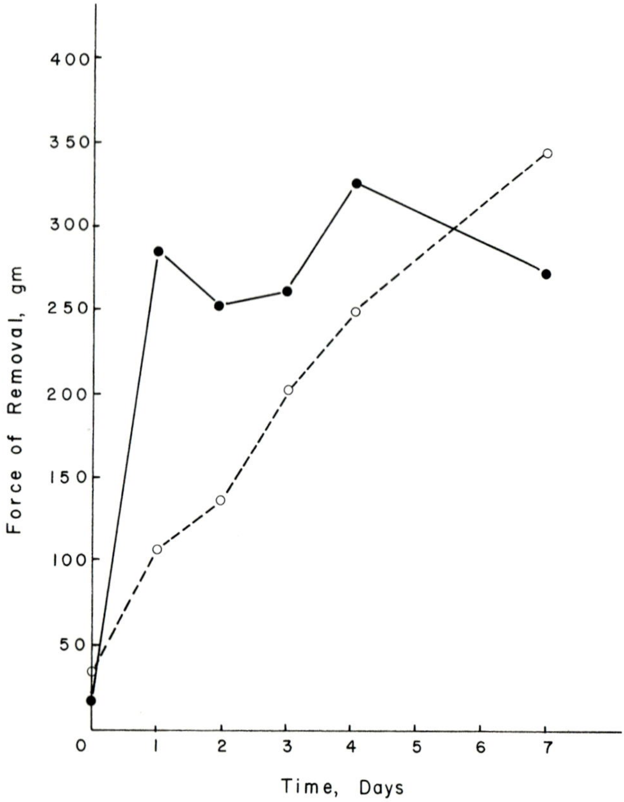

Fig. 4. Force of removal of occlusive (O---O) and nonocclusive (●——●) tapes removed and reapplied daily for 7 days (average of 5 subjects). The adhesion of the occlusive tape continued to build through the period indicating that lower, highly cohesive regions of the stratum corneum were being reached. The easily hydrated, upper layers were removed in the first 3 days. The nonocclusive tape maintained a rather consistent adhesion and removed only outer horny layers.

are found on the adhesive at any of the time periods. With the porous tape, therefore, the line of cleavage at removal develops high in the stratum corneum and less skin is removed.

IV. Discussion

Available evidence indicates that much of what has been called allergic reaction of skin to rubber based adhesive tapes may be attributed to traumatic rather than allergic response. These tapes are occlusive by nature of the hydrophobic adhesive, and when worn for more than a short period of time they allow moisture to build up in the outer, horny layers of skin. These outer layers at normal environmental humidities are dry and very strong, but when hydrated they become weak and easily damaged. Under these conditions the adhesion between the tape adhesive and the outermost layers is far greater than the internal strength of the stratum corneum. Irregular, deep fractures of these layers occur as the tape is removed, thus leaving large amounts of stratum corneum on the tape. This becomes exaggerated when fresh tape is reapplied over the same area. The subsequent redness that develops can often be confused with chemical irritation or "allergic" response.

Such traumatic response is minimized when surgical adhesive tapes are constructed so as to allow normal skin moisture to move through the adhesive and backing material into the environment. Acrylic adhesive with hydrophilic properties and a high moisture transmission, spread on a fabric or a nonwoven backing, represents a new generation of surgical tapes with low potential for trauma. Under these conditions the moisture content of the horny layers remains relatively normal and its strength is retained. When these porous tapes are removed, fracture lines develop near the surface in the region of the naturally desquamating layers and only small amounts of stratum corneum are removed. This allows repeated use of a tape, as in frequent dressing changes, with minimal trauma to the skin.

References

Goldschmidt, H., and Kligman, A. M. (1967). *Arch. Dermatol.* **96**, 572.

Jenkins, H. L., and Tresise, J. A. (1969). *J. Soc. Cosmetic Chemists* **20**, 451.

Kligman, A. M. (1964). *In* "The Epidermis" (W. Montagna and W. C. Lobitz, Jr., eds.), pp. 387-433. Academic Press, New York.

Orentreich, N., Berger, R. A., and Auerbach, R. (1966). *Arch. Dermatol.* **94**, 709.

Pinkus, H. (1951). *J. Invest. Dermatol.* **16**, 383.

Tregear, R. T., and Dirnhuber, P. (1962). *J. Invest. Dermatol.* **38**, 375.

Wolf, J. (1939). *Z. Mikroskop.-Anat. Forsch.* **46**, 170.

PART IV

ADHESIVES FOR HARD TISSUE

CHAPTER 15

BONDING TO HARD DENTAL TISSUES

Michael G. Buonocore

I. Potential Uses for Dental Adhesives

A. Adhesives as Restorative Materials

Materials currently employed in the restoration of teeth have a number of shortcomings, a major one being their inability to provide a leakproof marginal seal for the tooth cavities. The failure to seal cavity margins may be responsible for pulpal irritation and/or the recurrence of dental decay due to percolation of bacteria, acids, and other harmful oral products between the walls of the restoration and walls of the tooth cavity (Nelson *et al.*, 1952). Marginal percolation of harmful oral products is in large part due to differences in coefficients of thermal expansion between the restorative material and tooth structure. Thus, alternate exposure to hot and cold foods such as coffee and ice cream could result in alternately reducing and enlarging the size of the marginal space—creating a sort of pumping action by which oral debris is drawn in, pumped out, and drawn in again. To minimize the effects of marginal leakage these vulnerable marginal areas are placed by the dentist in self-cleansing regions of the tooth surface where they are more likely to be free of accumulated oral debris. This requires that the size of a dental cavity be extended laterally with the result that healthy tooth structure is removed along with the unhealthy—a wasteful, time-consuming, but presently necessary procedure. Figure 1 is a cross section of a molar tooth showing two decay areas which have penetrated both the enamel and dentin, the latter being both pigmented and otherwise altered in appearance for some depth. Treatment of this tooth requires that the decay area be removed. In Fig. 2 the shaded area represents the region where a restorative material would be placed after the removal of the decay and extension of the margins of the cavity laterally into self-cleansing areas. The cavity preparation is box shaped as usual, with the walls flaring somewhat at the base of the cavity to provide mechanical retention for the restorative material. If the restorative material were adhesive, the undercut, box shaped cavity would not be required. It could be replaced by a bowl shaped type as outlined by the shaded area in Fig. 3. Such modification would result in savings of time and tooth structure. Other cavity design modifications could involve only the removal of the immediate areas of decay shown in Fig. 1, resulting in two small cup shaped cavities with the intervening tooth structure left undisturbed. There would be little concern here about marginal percolation or leakage since the bonding of tooth structure and restorative material would have eliminated the space between them.

B. *Adhesives as Tooth Coatings to Prevent Dental Decay*

One of the areas most vulnerable to dental decay is the occlusal or biting surface of posterior teeth, an example of which is shown in Fig. 4. One notes here the presence of a large central fissure or groove from which multiple finer crevices radiate, some of which are deep enough to extend into the dentin. Decay starts in these fissures and crevices soon after the teeth erupt and enters the dentin to produce the decay lesions shown in Fig. 1. An adhesive material that could be applied

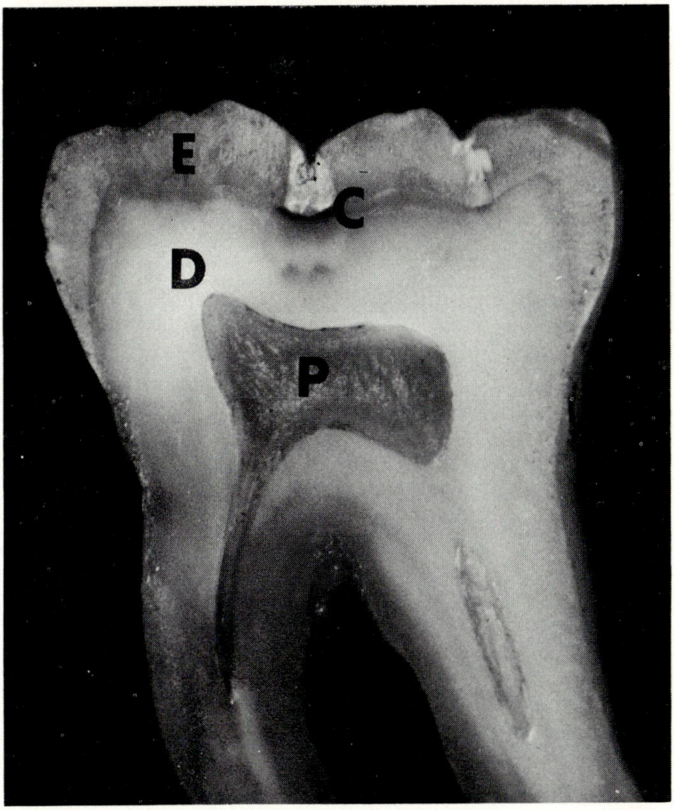

Fig. 1. A mesio-distal cross section of a tooth through the central fissure area. E = enamel; D = dentin; P = pulp; and C = central decay lesion showing typical whitish appearance of decalcified enamel. A second lesion is to the right. Note changes in the dentin as a result of decay, including the dark crescent shaped area at the junction of enamel and dentin.

rapidly and easily over the enamel surface of recently erupted teeth to shield these vulnerable pits and fissures from decay would be particularly beneficial for children.

C. Adhesive Cavity Liners and Cementing Materials

An alternate means of protecting teeth against marginal leakage would be to use a thin layer of adhesive over the floor and walls of a cavity before filling it with a restorative material. If the restorative material in turn also bonded strongly to the liner, marginal leakage would be eliminated. Even if the restorative material did not bond

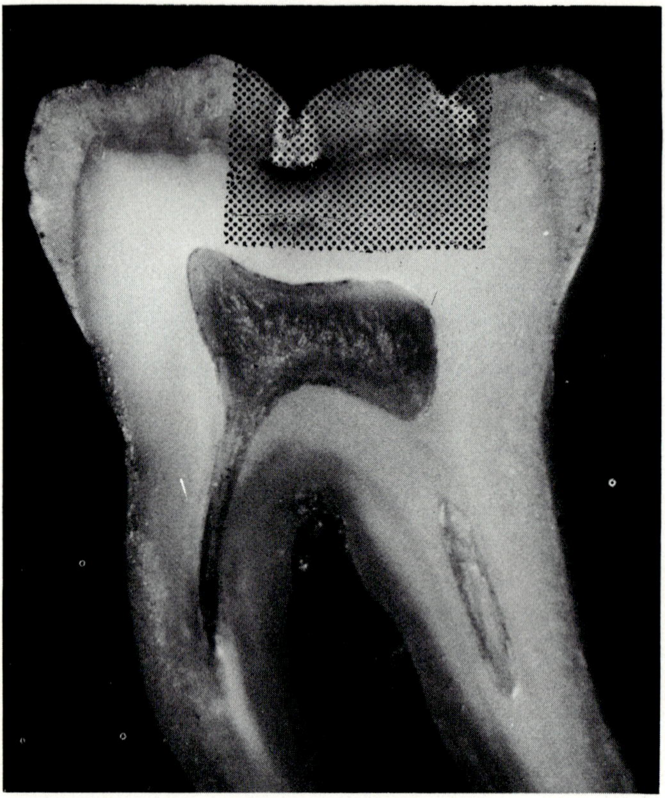

Fig. 2. The shaded area shows the extent of the cavity that is usually drilled in removing the decay. Note the slight outward flare of the cavity walls at base of the cavity to aid in retention of the restoration.

with the liner, the percolation of harmful oral products would nevertheless be limited to the space between the walls of the restoration and the liner. If the liner were resistant to degradation by these percolated oral products, it could protect the tooth by preventing contact with harmful products. Figure 5 shows diagrammatically the function of an adhesive cavity liner as described above.

The agents now used for cementing gold crowns, inlays, and orthodontic bands to teeth suffer the same limitations as other dental materials. The interface between the cement and tooth substance is subject to leakage with the result that recurrent decay and/or pulp irritation may occur under these materials. This may be particularly dangerous under metal crowns and orthodontic bands since the extent of the decay cannot be adequately diagnosed by X ray; X-ray passage through the tooth is blocked by these materials.

D. Adhesives for Other Uses

An adhesive for dental enamel would have many advantages.

1. Adhesives for Cementing Orthodontic Brackets Directly to Teeth

During orthodontic treatment, steel bands are usually placed around teeth and serve as a base to which brackets are welded to provide anchorage for wires. An adhesive suitable for bonding brackets

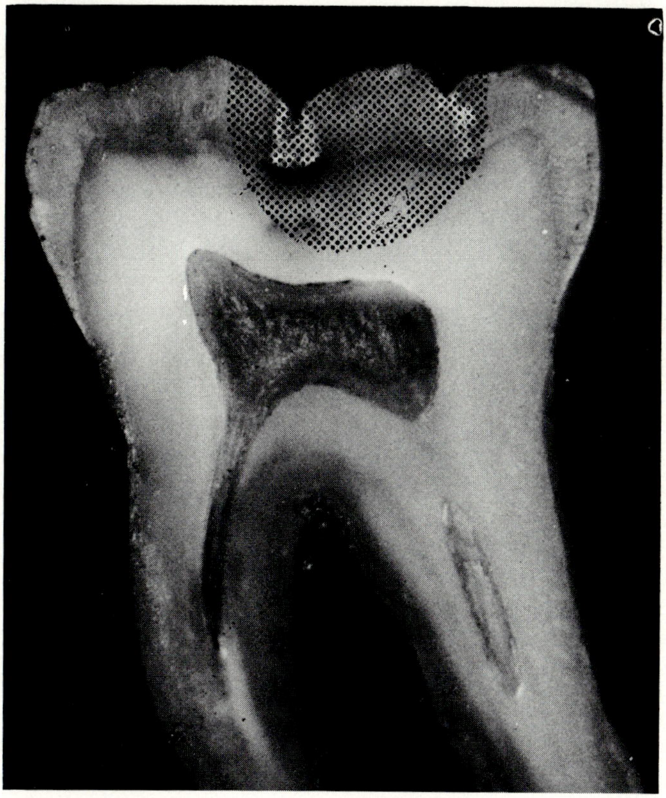

Fig. 3. The cup shaped nonretentive shaded area is a modified cavity preparation that could be used in place of that shown in Fig. 2 if the restorative material were adhesive.

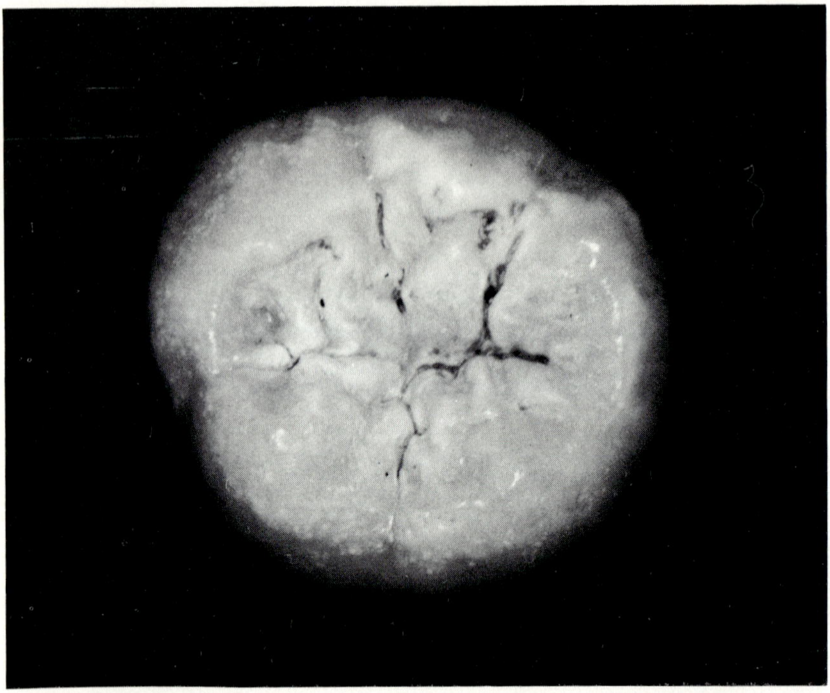

Fig. 4. Typical pits and fissures of the biting surface of a posterior tooth. Note the relatively wide, deep, and darkly stained fissures at the right. On the left the fissures are finer and not as deep. Decay in these fissures in cross section might appear as shown in Fig. 1.

directly to enamel surfaces would eliminate not only the need for time-consuming placement of unaesthetic bands but also, as a final step in treatment, the closing of the space between teeth created by the use of the steel bands.

2. Adhesives for Repairing Small Enamel Fractures of the Anterior Teeth

A common accident among young children is fracture of the biting edges of the front teeth producing an unaesthetic appearance which, for valid reasons, is often not corrected until the child is in his teens. An adhesive material would provide a rapid convenient means to repair such fractures immediately.

3. Adhesives for Covering Discolorations and Hypoplastic Blemishes on Surfaces of Anterior Teeth

In recent years the use of tetracyclines has been associated with unsightly types of discolored defects such as pits, depressions and/or ridges on the surfaces of the anterior teeth (Frankel and Hawes, 1964). An adherent, pigmented coating could cover such defects satisfactorily.

II. Factors Affecting the Establishment and Maintenance of Bonding *in Vivo*

A. The Oral Environment

1. Moisture and Variations in Temperature

Adhesion to hard tissues has more requirements than those encountered for industrial adhesives. Adhesion to body tissues involves an aqueous environment both before and after bonding—a situation that most adhesive chemists would consider detrimental to the establish-

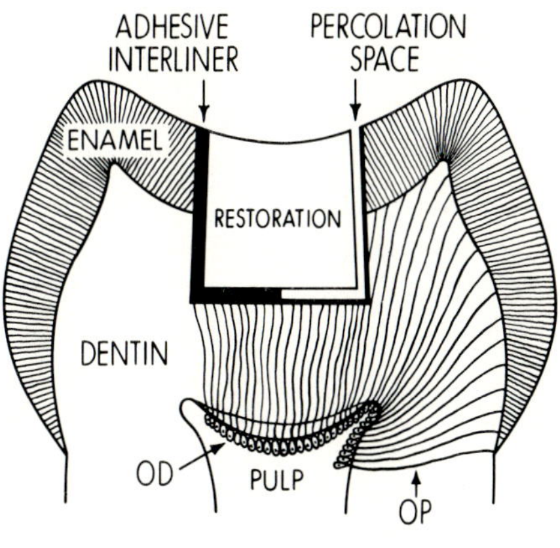

Fig. 5. A restorative material placed over a liner which is adhesive for the cavity walls. At the right side, the restorative material does not adhere to the liner; hence a percolation space due to thermal change may be produced. Provided the liner is resistant to degradation by percolated oral debris, the tooth will nevertheless be protected as the debris will be confined to the area between the restoration and liner. The left side shows a liner which is adhesive for both the restorative material and cavity surface. Because of its elasticity the liner stretches as the restorative material shrinks due to thermal change; the liner thus maintains the seal and percolation is prevented. OD = odontoblasts; and OP = odontoblastic processes.

ment and long-range performance of an adhesive bond. Water plays an important part in adhesion to any hydrophilic surface since it is strongly attracted to and held by such surfaces. This water must be displaced in order to obtain intimate contact between an adhesive and tooth structure. Moreover, to maintain the bond, the attraction between adhesive and tooth structure, if due to physical forces, must be equal to or stronger than the attraction of either phase for water and, if due to chemisorption or chemical bonding, must be resistant to hydrolysis. Wet conditions would impose serious limitations on the choice and performance of industrial adhesives and are avoided. For both medical and dental applications, however, there is no alternative; an adhesive must be capable of bonding to surfaces that may not tolerate complete drying and more important, the adhesive must maintain its adhesion in continuous contact with water. The bond between adhesive and tooth structure may also be subjected to significant sequential variations in temperature associated with eating. While these temperature variations might not be sufficiently great to impair the performance of a high bond strength industrial adhesive, they may eventually weaken or destroy oral bonding if it is not initially strong.

2. Oral Microorganisms

High concentrations and many varieties of bacteria and their metabolic products are present in human mouths. Many resins and resin adhesives have excellent resistance to bacterial degradation, but usually such materials are optimally cured at high temperatures and pressures, which conditions probably contribute to their resistance. Room temperature cured adhesives, of low exotherm, such as would be required for bonding to vital teeth, frequently have less than optimal properties and quite likely less resistance to bacterial or hydrolytic attack.

3. Masticatory Forces

The survival of adhesives to teeth is also lessened by shear, tensile, compressive, and torque forces of mastication. The success of a dental adhesive may well depend not only on its location in the mouth but also on its location in a tooth. Average biting forces are only a few pounds, but point contacts involving cusps, incisal edges, or unexpected hard particles may well bring about pressures of thousands of pounds per square inch, enough to induce failure of the adhesive (Skinner and Phillips, 1967). The small bonding areas on teeth contribute to disruption by relatively small forces.

B. The Tooth

Enamel and dentin are qualitatively similar but quantitatively quite different. Both may require different adhesives and methods of surface preparation and different techniques of application.

1. Enamel

This substance comprises a cap over the crown of the tooth covering the underlying vital dentin and pulp tissue (Fig. 1). Mature enamel is not changed in size or shape through physiological forces. It consists of about 95% hydroxyapatite with a calcium : phosphorus ratio of a little more than 2 : 1 by weight. The crystallites vary considerably in size (Scott, 1960; Scott and O'Neil, 1961; Trautz, 1967) and are considered to be surrounded and/or separated from each other by thin sheets or threads of a keratin-like protein material which comprises about 0.5% of the total weight of enamel (Hess, 1961; Helmecke, 1967). Loosely and firmly bound water accounts for roughly 4.0% by weight of the enamel (Brudevold and Söremark, 1967). Because of its external location on the tooth and its apparent lack of vitality, enamel can be dried more completely, as is done clinically with warm compressed air, than dentin and in this respect provides a potentially more desirable adherend surface. Since the organic matter of enamel is limited in amount (although its volume is large), bonding to enamel will probably occur through its inorganic fraction.

Although enamel is not considered a living tissue it nevertheless undergoes a variety of ionic substitutions, confined largely to its outermost layers, as a result of exposure to a variety of chemical agents derived from solid food and drink (Buonocore and Bibby, 1945; Brudevold *et al.*, 1960). Because of these interactions and also the acquisition of surface organic films from saliva and foods (Bibby and Van Huysen, 1933), enamel possesses a fully reacted, low energy surface not ideally suited for bonding (Zisman, 1961). It has been shown (Buonocore, 1955) that this fully reacted, relatively inert outer surface must be mechanically or chemically removed and/or altered in order to increase the water resistance of the bonding to enamel.

Bonding of restorative materials to cavity walls may fail because of the presence of structurally weakened and/or loosely attached enamel and dentin dust produced by cavity preparation. Structural weakness can scarcely be detected by the dentist, and removal of tooth dust is usually imperfect (Provenza and Sardana, 1965), but either defects will lead to bonding failure. The bonding strength of the joint will be limited by the strength of attachment of extraneous or weakly attached layers to the cavity surfaces. Even if a relatively clean surface is prepared, the presence of firmly bound water may still cause difficulty by adversely affecting the wetting characteristics of the adhesive.

2. DENTIN

The composition and structure of dentin has been extensively reviewed (Johansen, 1967; Bradford, 1967; Eastoe, 1967; Rowles, 1967). This tissue resembles bone, having approximately 20% organic matter which is largely collagen. The inorganic component comprises about 60% of the dentin and is hydroxyapatite of approximately the same chemical composition as that in enamel. The crystallites of dentin are much smaller than in enamel and are considered to be embedded in and completely surrounded by the organic matter. The remaining 20% of dentin is water. In contrast to enamel, dentin is a vital tissue having extensions of the pulp, called odontoblastic processes, traversing its substance radially. These vital extensions also transport tissue fluids from the pulp through the body of the dentin, a factor which may be of considerable importance in bonding to this tissue. Rigorous drying of dentin may cause injury to these protoplasmic extensions and hence to the pulp with the concomitant possibility of tooth loss (Langeland, 1957; Brännström, 1960; Cotton *et al.*, 1965). Because of its high organic content and comparatively loose structure, dentin is more permeable than enamel permitting the relatively rapid penetration of large molecules through its substance. This is particularly important from the aspect of penetration of toxic monomers from unpolymerized adhesive mixes applied to the surface of dentin. On the other hand, the larger amount of organic matter in dentin may offer a greater potential for bonding through chemical reaction with the adhesive. This possibility, however, is subject to the limitation that such reaction may produce adverse pulp responses.

The greater amount of water present in dentin is another factor militating against adhesion to this tissue. Even assuming that the surface layers of dentin could be dried sufficiently to obtain bonding with a restorative material, the bonded surface would be subjected very shortly to the possible deleterious effect of moisture resupplied by the tissue fluids passing from the pulp through the dentin via the odontoblastic processes.

These differences in composition between enamel and dentin appear to be important in determing adhesive behavior as shown by the rather strong bonding of a methylmethacrylate filling resin to acid conditioned enamel surfaces and the absence of bonding to similarly treated dentin surfaces (Buonocore, 1955). Another example of the divergent behavior of these tissues to the same adhesive was shown by a polymerizable formulation containing glycerophosphoric acid dimethacrylate which bonded much more strongly to dentin than to enamel (Buonocore *et al.*, 1956). These findings have important bearing on the performance of adhesive restorative materials since, when placed in a cavity, they may bond to one tissue and less strongly or not at all to the other, thus inducing undesirable stress relationships.

To eliminate the problems of marginal leakage and the need for undercutting to obtain mechanical retention of restorative materials, bonding to both enamel and dentin would be desirable. This would also provide a greater bonding area, producing a higher overall bonding strength. If a choice between the two tissues had to be made, bonding to enamel would be preferred. One might, however, question whether present cavity design offers sufficient enamel bonding area to maintain bonding under occlusal stress. To increase the contact area between enamel and adhesive, cavity design could be modified by substantially flaring the enamel walls as seen in Fig. 6a. This type of preparation would also serve to bring the margins of restorations into self-cleansing areas (Fig. 6). With a strong adhesive this additional bonding area might suffice to maintain the bonding under most oral situations. It can be seen by comparing part a with part b in Fig. 6 that adhesion to dentin alone would be less desirable. Although marginal leakage of dye molecules shown by the dark area between the enamel and the restoration might be effectively stopped at the bonded dentin-restoration interface, it could detour along the very permeable dentin-enamel junction to penetrate the dentin beyond the bonded interface and to reach the pulp by this less direct route (Buonocore, 1963).

C. The Adhesive

The properties of the adhesive, their interaction with, and their relationship to the adherend surface are fundamental to obtaining successful adhesion.

1. Polymerization Shrinkage

The polymerization of liquid monomers and the subsequent cooling of the polymerized mass is attended by loss of volume, which causes the adhesive to pull away from the walls of a dental cavity. If the adhesion of the resin to the cavity were not sufficiently strong to resist the tensile force with which the resin is being pulled away, the bond would fail. If on the other hand, the adhesion were stronger than the tensile forces tending to disrupt it, the bond between cavity walls and

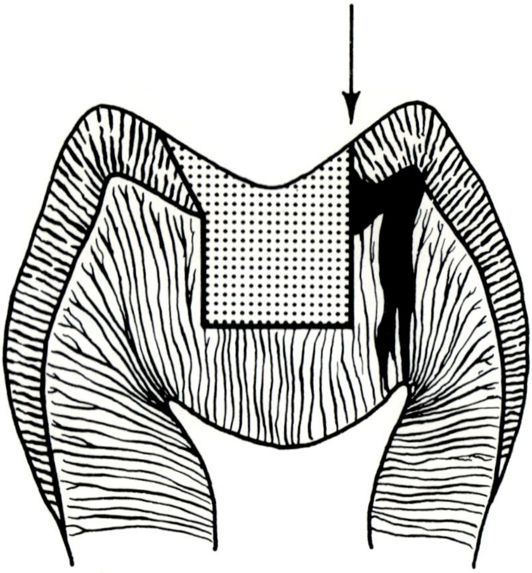

Fig. 6. Diagrammatic representation of a cross section of a restoration in a tooth cavity to demonstrate marginal leakage under different conditions of bonding. Part (a) shows the flaring of the enamel cavity wall to increase bonding area. With an adhesive restorative material no marginal leakage would occur even in the absence of bonding to dentin. Part (b) shows the leakage pattern of a dye where there is bonding of the adhesive to dentin but not to enamel. The dye in this case could penetrate (arrow) the enamel-restorative material interface until it reached the bonded dentin-restorative material interface where it would be stopped. However, it could then diffuse along the very permeable junction of enamel and dentin entering the latter via the odontoblastic processes.

restoration might survive although it would be under considerable stress. According to Bowen (1967), stresses of the order of 1280–1660 pounds per square inch may occur during the hardening of an unreinforced direct filling resin provided the bonding to the cavity walls does not fail. In these tests exposed surface area was equal to 20% of cavity wall area. Such a situation is not stable, and additional small forces may cause spontaneous rupture of the bond. Shear, tensile, or compressive forces are caused by mastication or by strains imposed as a result of thermal changes. Stresses associated with polymerization shrinkage and/or with cooling shrinkage following polymerization should be minimized or eliminated. This can be accomplished to some extent by mixing with the monomer high concentrations of inorganic fillers having low coefficients of thermal expansion. Better still would be the use of high molecular weight monomers of low polymerization shrinkage (Bowen, 1962), which are being used in certain restorative resin compositions (Buonocore, 1968). One such monomer is the adduct of the diglycidyl ether of bisphenol A and methacrylic acid, the chemical formula of which is shown below.

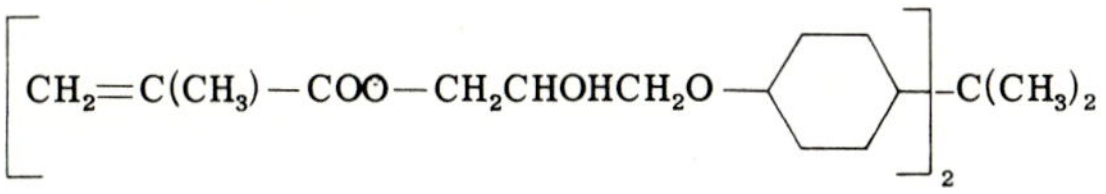

(1)

When this is combined with concentrations of inorganic filler such as powdered silica, and glass beads, which can comprise up to 70% of the total weight of the formulation, polymerization shrinkage is considerably reduced (Bowen, 1963).

Another means for reducing the stresses due to polymerization shrinkage would be through controlled expansion of the hardened resin as a result of hydration. Polar groups in the adhesive should be conducive to hydration, but control of expansion by this means will require extensive study. Stresses could also be reduced if the ratio of exposed free surface to bonded surface in the cavity were increased (Bowen, 1967), such as by widening the cavity without increasing its depth.

2. Polymerization Conditions

Because high temperatures will damage teeth, an adhesive should not generate excessive heat during polymerization. The use of high concentrations of fillers, by reducing the amount of polymerizable monomer present in an adhesive mix, would reduce the total heat of polymerization. A similar effect could be obtained if high molecular weight monomers or partial polymers were used. Here the ratio of polymerizable group length to total chain length is small compared to an equal weight of a low molecular weight molecule containing the same polymerizable group. High molecular weight viscous monomers also penetrate by capillary action less rapidly into dentin than the free flowing low molecular weight volatile fluids such as methylmethacrylate monomer, commonly used in self-curing restorative resins. If the monomer and/or the polymerization catalysts and accelerators usually dissolved in it were strong irritants, extensive pulp damage would occur. Moreover, by the introduction of fillers the high molecular weight materials can be formulated as thick, relatively dry pastes. When these are blended prior to use with small amounts of polymerization catalysts, the mixture can be packed into a cavity with much less risk that appreciable amounts of free monomers will penetrate the dentin before polymerization occurs. Unfortunately, the relatively dry, pasty adhesive mix may not be conducive to adequate wetting of the tooth surface to establish maximum adhesion.

3. Physical and Chemical Properties

a. Coefficients of Thermal Expansion. A major factor militating against maintenance of adhesion is the difference between coefficients of thermal expansion of adherend and adhesive. Teeth are subject to relatively wide extremes of temperatures. One reason for bond failure of self-curing acrylic resin restorative materials may be that they expand or contract seven or more times as much as tooth structure for each centigrade degree of temperature change (Skinner and Phillips, 1967), and in the wet state acrylic resins may have considerably greater coefficients of thermal expansion (Spreter von Kreudenstein, 1953). Considerable stress will be placed on a bond between an adhesive restorative material and tooth structure if such disparities in coefficients of thermal expansion exist. Efforts to match coefficients of thermal expansion have involved the use of large concentrations of inorganic fillers, preferably those that have negative coefficients of thermal expansion such as β-eucryptite, a lithium aluminum silicate (Gillery and Bush, 1959).

b. Polar Groups. The hydrophilic nature of tooth structure would be best matched by the use of adhesives containing polar groups. The polar groups would not only provide for internal bonding among polymer chains to produce a strong material, but also contribute to better adhesion. However, water absorption might also be favored by the presence of polar groups and could result in excessive swelling and/or softening of the adhesive with possible concommitant weakening or loss of bond strength (Hoekstra and Fritzius, 1951).

c. Toxicity. The toxicity of materials used as hard tissue adhesives is of primary concern. Distinction, however, should be made between the toxicity of the monomer and that of its polymer. A fluid mixture of a potentially toxic monomer in contact with vital dentin for only a short time before polymerization may not be harmful. Hence such monomers should not be excluded from evaluation as potential adhesives. However, a toxic or nontoxic monomer may produce a polymer which is toxic by virtue of a slow release of harmful products resulting from gradual reaction with and degradation by the oral environment. Thus, the evaluation of adhesives to be used as restorative materials must involve long-term studies because pulpal and other reactions may not be apparent over short periods. In addition, since restorative materials are in direct contact with other mouth tissues, observations should also be made of the long-range effects of such contact.

d. Color. Poor color in adhesive materials, while having little to do with their ability to bond, would eliminate them from consideration as potential restoratives for anterior teeth because of aesthetic considerations. For the posterior teeth, however, the disadvantage of poor color would be far outweighed by the benefits to be derived from the ability to form leakproof bonds with tooth structure. Evaluation of adhesives should be considered regardless of their color since information may be obtained on elements of structure and composition favorable for adhesion.

e. Factors Related to Ageing. Under this category one can consider a variety of factors which may affect the maintenance of adhesion to oral structures. Long-term observation of adhesives may be required under simulated *in vivo* conditions, to ascertain such changes as bond failure, stress fatigue, crazing, solubility, swelling, bacterial degradation, softening, and discoloration.

III. Approaches to Bonding to Teeth

The approaches used to obtain bonding to tooth structure have been directed at modifications of the tooth surface, the synthesis of new monomers specifically tailored for adhesion, and the investigation of biological adhesives.

A. Modifications of Hard Tissue Surfaces for Enhancing Adhesion

1. Acid Conditioning of Hard Tissue Surfaces

In industry the proper preparation of the adherend surface is essential for optimal bonding. Surface preparation has to take into consideration the nature of the adherend and type of impurities or surface contaminants likely to be present. It may range from the removal of loosely bound dust by vacuum to elaborate techniques for oxidizing surfaces of polyethylene to increase printability. A method which has been employed successfully for conditioning metal surfaces uses a resin formulation containing phosphoric acid. It is of particular relevance to the dental situation in that the surface conditioning was designed to improve the adhesion of paint and resin coatings to metals for under-water service (Quarles, 1953). The success of this wash primer and similar pretreatments represents a means of bridging the contrast in properties of inorganic materials and synthetic organic resins, and it has spurred the investigation of phosphoric and other acids as a means of conditioning enamel surfaces to improve bonding.

Since the surface of tooth enamel has reacted with various ions and molecules derived from saliva and foods to which it has been exposed, and its tiny imperfections become filled by a variety of adventitious materials, the composition of the superficial surface is quite different from that of the underlying enamel (Buonocore and Bibby, 1945; Brudevold *et al.*, 1960). As a result, the enamel has lost any receptivity to adhesion which it originally may have had. To render enamel more receptive to adhesion it was etched by a variety of acids including hydrochloric and phosphoric acids and a combination of a phosphomolybdate reagent (Folin-Wü) with oxalic acid. While all of the acids tested gave substantially higher strengths and duration of bonding under water than did unetched surfaces, the best overall results were produced by the use of a few drops of an 85% solution of phosphoric acid applied to the enamel surface for about 30 seconds. This acid was additionally favored because it produced less decalcification than the stronger mineral acids (Gwinnett and Buonocore, 1965). After pretreatment the acid was thoroughly washed from the surface and the enamel dried with warm compressed air for several seconds before application of a self-curing resin restorative material as the adhesive. The results of tests with the above technique using anterior teeth *in vivo* showed a minimum of a 100-fold increase in duration of bonding compared to unetched enamel surfaces (Buonocore, 1955). The strength of the bonding for test specimens which had been in the mouth up to 2 months was so great that attempts to remove the adhesive from the surfaces with a sharp dental chisel frequently caused fracture of the adhesive rather than clean separation from the surface. This indicated that the bond between enamel and adhesive was stronger than the adhesive itself. After longer periods in the mouth, as much as one year in some instances, less force was required to produce clean separation of the adhesive from the enamel surface. After removal of the adhesive, the enamel surface was somewhat whitish and dull but returned to normal appearance within a few days.

The increase in bonding strength attributable to phosphoric acid may be caused by a number of factors among which are (a) a tremendous increase in the surface area of the enamel due to the acid etching, (b) the exposing of the organic framework of enamel about and to which the adhesive can adhere, (c) the removal of old, fully reacted, inert enamel and exposure of a fresh reactive surface more favorable for adhesion, and (d) the exposure of a surface of an adsorbed layer of highly polar phosphate groups derived from the acid. To account for the increase in bonding strength observed with the phosphomolybdate-oxalic acid treatment it was suggested that a new receptive chemical surface was formed consisting of a bound complex of calcium oxalate and/or phosphomolybdate.

To obtain additional information on the factors influencing bonding to hard tissue the use of acidic surface treatments has been extended to dentin. A polymerizable resin mixture containing glycerophosphoric acid dimethacrylate was used as an intermediate adhesive layer between a dentin surface and a self-curing acrylic resin (Buonocore and Quigley, 1958). Dentin was tested untreated and treated for 1 minute with 7.0% hydrochloric acid. Tensile bonding strengths to untreated dentin were about half that found on treated surfaces and of shorter duration. As seen in Fig. 7, bonding values on acid-conditioned surfaces gradually decreased, reaching a plateau at the fourth

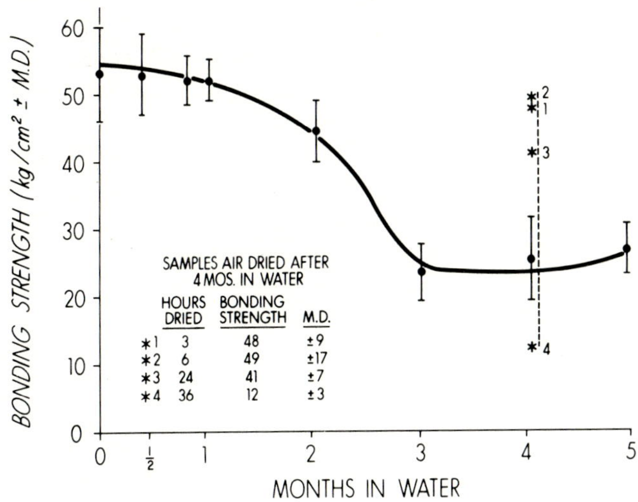

Fig. 7. Changes in bonding strength measured in tension of phosphate containing resin to acid treated human dentin following water storage and after drying for 3, 6, 24, and 36 hours following the 4-month immersion period. Each point represents an average of 15 tests, in tensile.

and fifth months which represented approximately a 50% decrease in bond strength from initial values. Samples showing tensile strengths of about 800-900 pounds per square inch generally showed portions of the adhesive still attached to the dentin. As occurred with enamel, bond strength decreased with time and later there was clean separation of the adhesive from the dentin surface.

The improvement in bonding strength resulting from acid etching of the dentin surface can probably be attributed to an increase in porosity of the surface, allowing for greater mechanical retention of the adhesive compared to untreated dentin. On the other hand, the probability must be considered that the increase in bonding strength could be due to chemical reaction between reactive groups in the adhesive and the organic matter of dentin. It might be speculated that if the initial bonding were largely due to mechanical retention, prolonged water immersion might progressively soften the phases so that separation from undercut areas would be easier. This would account for the lower bonding strengths observed with longer water immersion times. If this were the case, sufficient drying of the samples prior to testing might restore the yield values of both dentin and adhesive through rehardening and, thus, increase bonding strength by making undercut areas more retentive. If, however, bonding were due initially to chemisorption, the partial separation of the phases probably due to hydrolysis on prolonged water immersion would be expected to be permanent. Drying in this instance would serve merely to remove interfacial water leaving a hardened adhesive which would not be expected to reform a bond with dentin.

The results of progressively drying bonded assemblies which had been stored in water for 4 months are shown by the vertical dashed line in Fig. 7. It can be seen that after 3 and 6 hours drying the bonding strengths are restored to near initial values. Even after 24 hours drying there is considerable enhancement of bond strength over the "wet" 4-month values. The reduced bonding observed after 36 hours drying may be attributed to weakening and straining of the bond due to warpage and excessive shrinkage of both phases as a result of extreme drying.

While the results of these studies support the idea that mechanical interlocking is most likely the principal basis of the observed bonding strength, the possibility of concommitant chemical bonding between adhesive and enamel and dentin cannot be ignored. With dentin the latter idea is supported by the evidence of a surface zone of altered staining of the dentin in contact with the adhesive (orthochromatic with toluidine blue, while the remainder of the dentin stains metachromatically). The zone of altered staining appears to be associated with the presence of glycerophosphoric acid dimethacrylate (Kramer and McLean, 1952; Buonocore and Quigley, 1958).

With respect to enamel, recent evidence has shown that bonding to acid-treated surfaces may also involve mechanical interlocking. Comblike tags are present which appear to be attached to a self-curing resin restorative material after the enamel to which it is bonded is dissolved by acid (Gwinnett and Buonocore, 1965; Buonocore *et al.*, 1968). Figure 8 shows the appearance of such tags, which usually possess morphological characteristics of the enamel prisms. Acid etching of the enamel surface creates spaces between prisms into which the adhesive penetrates. When the resin penetrates the enamel prior to polymerization, it may also entrap and/or chemically combine with the organic matter present in the enamel spaces. Hence we believe the tags represent a combination of resin and organic matter of the enamel. That chemical bonding in addition to mechanical interlocking may exist with enamel is supported by evidence (Buonocore *et al.*, 1968) that the interface between enamel and adhesive is resistant to penetration by sulfate ion containing ^{35}S, even after the bonded assembly has been stored in water for 6 months. Figure 9 is an autoradiograph made from a section of a tooth self-curing resin assembly which

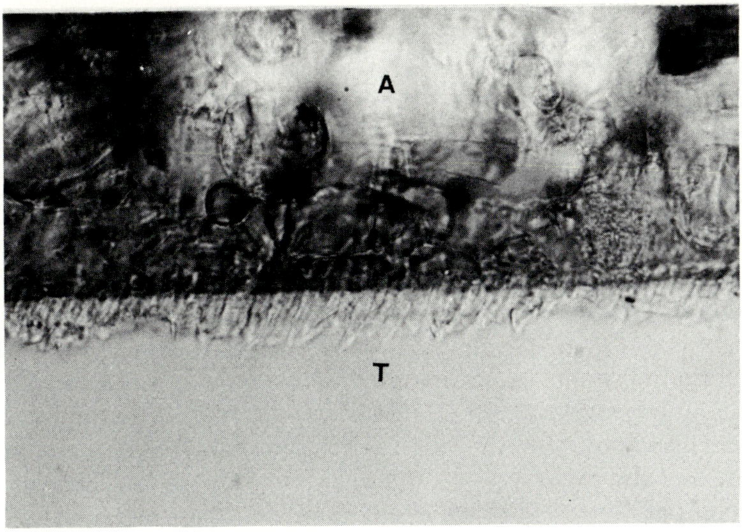

Fig. 8. Comblike tags (T) resembling prism outlines attached to the edge of an adhesive (A) that has been bonded to an enamel surface. Tags remain after enamel has been dissolved by acid.

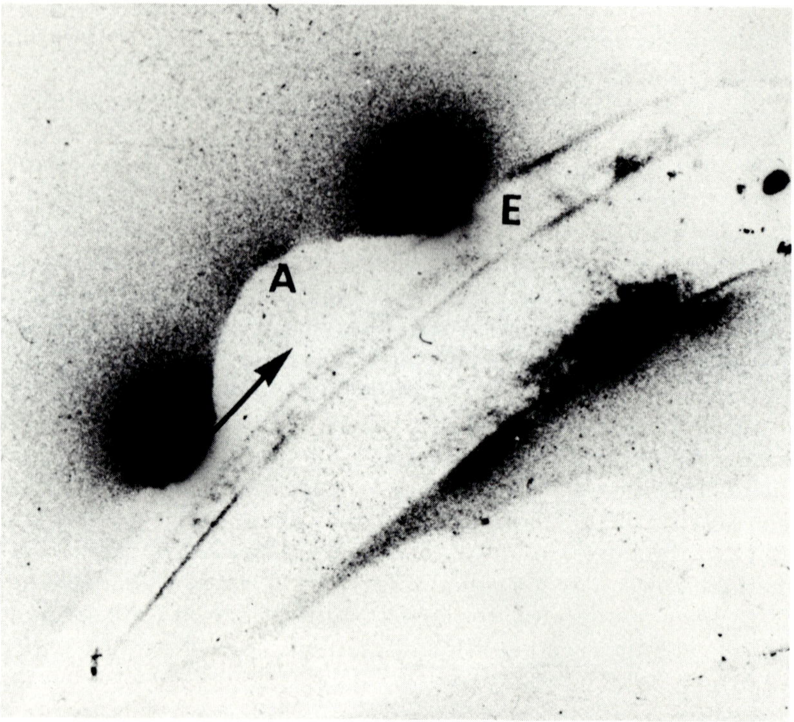

Fig. 9. Autoradiograph of a cross section of an adhesive (A) bonded to the surface enamel (E) of a bovine tooth. The adhesive-tooth assembly has been stored in water for 6 months. The enamel areas on each side of the adhesive which appear dark were exposed to a radioisotope solution for 24 hours. The path of isotope penetration through the enamel to reach the dentin is revealed by black regions. The interface (arrow) between enamel and adhesive, however, shows very little or no penetration of radioisotope and, thus, attests to the intimate nature of the bonding.

had been exposed to a radioisotope solution for 24 hours prior to sectioning of the tooth through the resin and underlying enamel. Just before radioisotope exposure, the tooth-resin assembly had been coated with wax, except for a 1-2 mm peripheral area surrounding the resin. This exposure to the isotope was thus confined to the bond interface and the peripheral enamel adjacent to the resin. Note the absence of penetration of radioisotope at the interface between the acid conditioned enamel and a self-curing resin restorative material. The black areas show the path of the radioisotope which has penetrated the exposed enamel and reached the dentin. This indicates that the body of the enamel is more permeable to the isotope than is the interface between resin and enamel—a degree of intimacy strongly suggestive of chemical bonding.

2. SURFACE ACTIVE COMONOMERS

Removal of surface is one means of altering hard tooth structure; another involves adding to it to produce a new surface capable of polymerizing with and/or bonding to the restorative material.

Compounds that might be useful in altering the surface should preferably have on one end of the molecule one or more polar groups that are capable of forming a strong bond to tooth structure and at the other end one or more nonpolar groups capable of copolymerizing with the restorative material. A compound having these general characteristics has been synthesized and called a surface active comonomer (Bowen, 1965a). It is the adduct of *N*-phenyl glycine and glycidylmethacrylate having the formula

$$HOOCCH_2-N(C_6H_5)CH_2CHOHCH_2OOCC(CH_3)=CH_2 \quad (2)$$

The *N*-phenyl glycine portion may be considered to form a chelate ring with calcium, thus bonding the molecule to the enamel or dentin surface. The methacrylate residue in turn would copolymerize with self-curing acrylic resins to bond them to the tooth surface.

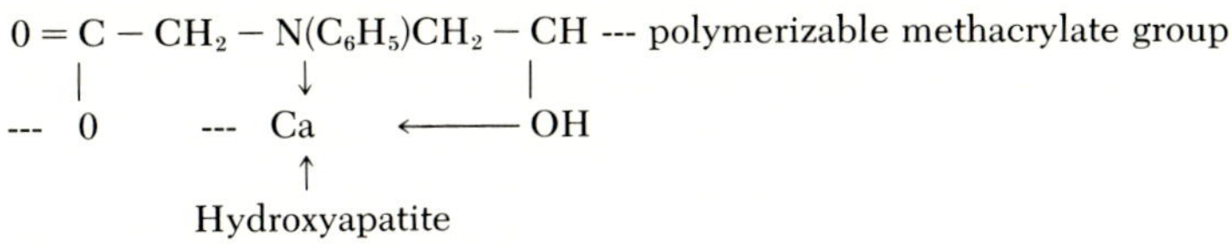

(3)

Reaction of the comonomer with the dentin or enamel surface is apparently with the inorganic rather than the organic phase. This is supported by the observation that enhanced bonding was also obtained on natural fluorapatite crystals which do not contain organic matter. Application of a 5 or 10% solution of this comonomer in ethanol or acetone, respectively, to dentin and enamel surfaces prior to application of a direct filling resin resulted in considerably higher bonding values than observed on untreated surfaces. In related experiments it was also observed that pretreatment of the dentin surface with dilute HCl or NaOH (Bowen, 1965b) prior to application of the comonomer further enhanced bonding strengths. To be effective as an intermediate bonding agent in the presence of water, the reactive group of the surface active comonomer should be capable of successfully competing with water for bonding sites on the hard tissue surface. Bowen (1961) has found that certain chemical groups appear to have greater affinity for hard calcified tissues than has water and that the relationship is dependent on pH. The increasing affinity of the groups tested was: hydroxyl, amino, carboxyl, and those capable of forming 5-member chelate rings with calcium. The affinity of water for dentin and enamel was considered to be between that of the hydroxyl and amino groups. Nitro groups have been suggested (Zettlemoyer, 1961) as capable of successfully competing with water. It follows that the greater the affinity of a particular chemical group for a surface, compared to water, the longer the bond will survive in the presence of water. Although surface treatment with comonomers has given improved bonding, the duration of the bonding compared to the service life of a restoration *in vivo* has not been adequately assessed. The reaction of the vital pulp to the application of surface active comonomers and/or their solvents must eventually be evaluated. In this regard neutral or near neutral solutions would be best suited for application to vital dentin surfaces particularly if there is only a thin layer of dentin overlying the pulp as in deep cavities.

3. Cavity Lining Materials

a. Commercial Film Forming Materials. The use of surface active comonomers probably involves "monolayers" of material deposited on a surface. This contrasts with existing dental cavity lining materials which are deposited from solutions of volatile solvents in the form of thin, varnishlike, visible films over the cavity surface. Cavity liners, which are natural or synthetic resins, were designed to seal off the dentinal tubules from ingress of harmful agents from restorative materials and oral debris (Swartz and Phillips, 1960). While theoretically the use of cavity liners offers greater potential for successful bonding than bulk material, no commercial film-forming materials have been found that will bond to enamel and dentin for any appreciable time when exposed to water; a number of such materials have been evaluated by Buonocore and associates (1962). Neither have available dental liners shown the capacity to prevent marginal leakage, although some reduction in leakage has been demonstrated (Gillings *et al.*, 1960; Going and Massler, 1961; Pinto and Buonocore, 1963). Further research is needed in this field since adhesive materials used as liners would circumvent one of the difficulties associated with their use in bulk in that they appear to give optimal bonding strengths in thin layers. This is in part attributable to the fact that thin layers of adhesive contain fewer voids and other imperfections than bulk material and are easier to attain in practice. In addition, shrinkage due to polymerization or as a result of solvent loss could be better tolerated, since the resulting tangential stresses acting at the bond interface would be smaller than in bulk specimens (Hoekstra and Fritzius, 1951). Moreover, a thin layer of material would permit rapid evaporation of solvent, the retention of which, occurring in a greater bulk of material, might impair its properties. Lining materials could be elastomers and serve adequately as a load absorbing and distributing medium, although they might not be usable as bulk restorative materials. Thin films would also allow for the quick dissipation of heat from exothermal reactions and would broaden the scope of materials that could be employed as adhesives.

b. Liners from Polyisocyanates. Potential lining materials produced from reactive polymers or prepolymers such as the polyisocyanates have received attention in recent years (Galligan *et al.*, 1968), and are discussed in detail elsewhere in this volume. The isocyanate group is capable of reacting quickly with hydroxyl, amino, and carboxyl groups to form relatively stable urethane linkages. The presence of protein in enamel and particularly in dentin, where it occurs in large amounts, therefore, provides an ideal substrate for reaction with isocyanates to form a chemically bound polymeric surface over the hard tissues. Because of their physical and chemical properties, the isocyanate prepolymers would best be employed as comparatively thick coatings, perhaps up to 100 μ, to which resin adhesives may be bonded. This would reduce the risk of dissolving the liner as is the case with existing dental cavity liners if one attempts to place self-curing resins over them. While the polyisocyanates appear to offer considerable promise as lining materials or bases, their high reactivity, particularly to water, may be a disadvantage. This is so because the reaction of the isocyanate group with water produces carbon dioxide which may cause foaming of the polyisocyanate liner. Unpublished studies (Buonocore, 1964), in which this type of liner was applied over air-dried dentin, showed that foaming occurred and continued over a period of time following placement of the restoration. This and subsequent swelling of the liner on prolonged water immersion caused displacement of the overlying restoration in many instances. The incorporation of fillers such as calcium hydroxide, capable of adsorbing or reacting with carbon dioxide, might in part circumvent this problem.

A recent report (Buonocore and Casciani, 1969) outlined the synthesis of polyurethanes from the reaction of various high and low molecular weight mono- and polyisocyanates with hydroxyethylmethacrylate. The addition reaction products of this hydroxy compound with DuPont Co. "Adiprene" isocyanate prepolymers L100, L167,

L213 and L315 gave viscous yellow liquids or semisolids, which when polymerized in cavities *in vitro* produced hard tough products which tended unfortunately to soften somewhat on water immersion. Marginal leakage tests, however, with these addition products revealed almost complete absence of marginal dye penetration both in noncycled and thermally cycled specimens. These results are the best that have been obtained in our laboratory in comparison with a considerable number of conventional and experimental restorative materials. In view of their toughness and resiliency in addition to other favorable properties, these high molecular weight polymerizable compounds may have excellent potential as cavity lining materials.

B. Compounds Specifically Tailored for Adhesion

1. Polyacrylic Acids

A more recent approach to obtaining adhesion to teeth is the development of a polycarboxylate composition capable of bonding to dentin and enamel (Smith, 1968). The liquid component is a 25% water solution of a polyacrylic acid; the solid phase consists of a specially treated, very finely powdered zinc oxide containing about 10% magnesium oxide. When the proper proportions of these two ingredients are mixed, a smooth, heavy consistency paste is obtained that soon changes to a hard, brittle, light pink opaque solid. Bonding is obtained when the mixture is applied as soon as possible after mixing to enamel which has been previously cleaned by brush and pumice or to freshly cut, air dried dentin surfaces. Tensile bonding strengths approaching 1500 pounds per square inch have been reported (Smith, 1968). These tensile strength values are, in effect, measures of the strength of the material itself since the adhesive bond is not broken. The material also appears to bond to properly cleaned metal surfaces such as orthodontic bands. Compressive and tensile strength of the polyacrylate materials compare favorably with those of zinc phosphate cement.

The chemistry of the polycarboxylate composition is such as to provide a large number of polar carboxyl groups for combination with the inorganic phase of hard tissues. In fact the lower bonding values obtained with dentin compared to enamel suggests that the organic matter of dentin may interfere in the establishment of adhesion. The polyacrylic acid is a polymer of acrylic acid with a molecular weight of about 20,000. Its water solution has a low pH and high titratable acidity which could cause pulp damage if it were applied in this condition over dentin. The combination with zinc oxide, however, results in the neutralization of the carboxyl groups with the formation of an insoluble zinc salt of the polycarboxylic acid. If complete neutralization of carboxyl groups occurs before the mixture is applied to the enamel or dentin surface, bonding is not likely to result. Hence mixing proportions should be adjusted so that residual carboxyl groups are available for bonding to enamel or dentin. It is probably best to apply the mix to surfaces immediately after incorporation of the zinc oxide into the liquid to take advantage of a greater availability of carboxyl groups for bonding.

The carboxylate material is relatively new and while preliminary data indicate a good pulp tolerance when it is employed as a base or cementing medium, additional study is needed before this material can be given general clinical acceptance. The carboxylate material may have potential for cementing orthodontic bands and brackets to teeth, and possibly for sealing pits and fissures of teeth for caries prevention. Unfortunately the use of carboxylate cement as a restorative material for anterior teeth does not seem to be feasible since it is quite opaque, lacking the translucency required in these situations. Its use in posterior teeth may be feasible. Research should be directed to synthesis of chemically related compounds possibly having other highly polar groups in the molecule or a combination of such groups along with the carboxyl. One decided advantage of the polyacrylic acid is the fact that it is water soluble, thus eliminating the need for solvents such as alcohol and organic monomers, which are potentially dangerous to the vital pulp. The hydrophilic nature of water soluble materials containing strong polar groups would also tend to make them more suitable as an adhesive for tooth structure, itself a polar substance.

2. Barnacle Cements

A very recent approach to obtaining hard tissue adhesives involves the investigation of barnacle cements (Carderelli, 1968). Acorn barnacles stick well to most surfaces, even including polyfluorocarbons such as Teflon to which they not only bond but maintain the attachment under water. The conditions under which barnacles bond to most substrates bear a close parallel to those under which the body hard tissues exist. As judged from his analysis of limited amounts of the set cement, the composition indicates a polysaccharide containing a small amount of lipid component The set cement is virtually insoluble in water and appears to be unaffected by nitric, sulfuric, and hydrochloric acids and aqua regia whether these are diluted, concentrated, hot, or cold. Organic solvents except hot glacial acetic acid are ineffective in dissolving the cured cement. The average minimal bonding strengths measured in tension to stainless steel, which are at best questionable because of difficulty associated with the testing technique, appear to be on the order of 22 pounds per square inch for *B. eburneus* and *B. trigonus*. These bonding values are certainly not high and even if increased by a factor of ten, to compensate for the absence of complete adhesive coverage of the shell bottom, do not appear to be impressive for service in the oral cavity, except possibly where minimal masticatory stresses are involved. The fact that the barnacle cement is extremely sensitive to water, hardening rapidly on contact with it, appears to impose rather formidable storage problems in the event that such a material is obtained in quantity. In addition, the resistance of the cement to attack by oral bacteria and the response of the vital pulp has yet to be evaluated. These tests cannot be made until a sufficient sample of the unset adhesive has been obtained—a feat which has so far defied solution.

IV. Clinical Applications of Adhesives

While there has been considerable research directed at developing hard tissue adhesives, clinical application has been delayed. Aside from a few pilot studies devoted to the problems of bonding orthodontic brackets to teeth (Swanson and Beck, 1960; Newman, 1965; Relief and Dreyer, 1967) and two broad clinical studies (Takeuchi *et al.*, 1966; Cueto and Buonocore, 1967) involving the sealing of pits and fissures of teeth for caries prevention, no other work of a practical clinical nature has been reported. Attempts to cement orthodontic brackets to teeth using cyanoacrylate adhesives have not been promising; bonding and reliability in service have been poor.

Sealing of Pits and Fissures for Caries Prevention

Adhesives have been used for preventing dental decay by sealing the pits and fissures of teeth; such treatment may be feasible as a routine adjunct to oral prophylaxis. The technique of sealing pits and fissures (Cueto and Buonocore, 1967) involved the conditioning of the enamel surface for adhesion by rubbing on enamel surfaces for 45 seconds, a drop or two of a 50% phosphoric acid solution containing 7% dissolved zinc oxide by weight. The teeth were then thoroughly washed and dried with warm compressed air for several seconds prior to application of the adhesive. The adhesive employed was a two-component system consisting of a free flowing colorless liquid, methyl 2-cyanoacrylate, also known as EK 910 prepared by Tennessee

Eastman Company, Kingsport, Tennessee, and a powder filler containing equal proportions by weight of each of the following: (1) silicic acid; (2) finished silicate powder; (3) silica gel; and (4) methylmethacrylate polymer powder. Components (1) and (2) of the filler are the active ingredients which cause the rapid polymerization of the monomer, whereas components (3) and (4) are the inert ingredients which, when intimately mixed with the active ingredients, attentuate the rate at which they polymerize the methyl-2-cyanoacrylate monomer. When the water content of the filler was about 9.0% (as determined by weight loss at 105°C for 12 hours), polymerization gave a stonelike, tooth colored polymer. The powder and liquid were mixed in a 1:1 ratio by weight to form a thick but limpid paste which was applied to the conditioned and air-dried tooth surface by gently tamping with a small bladed instrument and allowed to set. Excess set adhesive had to be removed by dental burs to eliminate high spots in biting which could cause tooth trauma. After treatment, patients were examined at 6-month intervals up to 2 years and records kept as to the extent of decay in control and experimental teeth, incidence of adhesive loss, and condition of the adhesive. The results after 1 year showed that the teeth covered with adhesive had 86.3% less decay of the pits and fissures than the control teeth. The reduction was obtained in spite of the fact that about 20% of the treated teeth had become either partially or completely uncovered sometime during the first six months. Moreover the adhesive that was retained tended to become reduced in amount either through abrasion, dissolution, or hydrolysis, or a combination of these factors. Because of these limitations the adhesive was replaced or added to at each six-month observation period, a situation not particularly desirable for clinical practice. The results of the two-year evaluation was similar to the 1-year study confirming that adhesive sealing by shielding the tooth from oral products can markedly reduce dental decay of pits and fissures (Ripa *et al.*, 1966). The results obtained by Takeuchi *et al.* (1966) were similar to those of Cueto and Buonocore (1967), although their techniques did not involve acid conditioning nor premixing of solid-liquid components. They also used methyl cyanoacrylate monomer.

Because of the great potential of adhesive sealing, the search for an improved adhesive composition and a simplified technique for its application was intensified. This search has resulted in the development of a new simple, rapid technique wherein a catalyzed liquid adhesive is painted on acid conditioned enamel surfaces with a fine brush. The use of a brush permits more precise control not only of the amount of adhesive applied but also of the extent of coverage. As a result no time-consuming occlusal adjustment is required to remove high spots. When hardened by exposure to long wave, ultraviolet light, the adhesive presents an almost transparent, smooth, glass-like surface which accumulates considerably less oral debris than uncovered pits and fissures. Fluoride compounds may also be incorporated in the adhesive and could afford additional long-range protection against caries by providing a slowly leachable source of fluoride ion that can be acquired by the teeth. Preliminary analysis of the data, which involves 200 sealed pits and fissures of both permanent and deciduous teeth, indicates that the adhesive will remain firmly bonded to the teeth for at least one year with little or no loss in volume of the original material and provides complete protection against pit and fissure decay (Buonocore, 1970). Compared to the earlier technique, which required reapplication of the adhesive composition every six months, the present method is more practical for clinical use. The greater simplicity and rapidity of this new treatment procedure, moreover, make it particularly appealing as a potential preventive measure which could be carried out by dental hygienists.

V. Concluding Comments

In discussing the need for adhesives, the properties of adhesives, and the problems associated with bonding, we have, in essence, outlined what we would like an adhesive to do. It is perhaps appropriate now to speculate whether an adhesive can be found that can do these things. We might begin by considering what an adhesive cannot do. An adhesive will not alter the bulk mechanical, physical, and chemical properties of enamel and dentin even though it may itself be superior in these respects. A bonded assembly is as strong as its weakest link. The tensile strengths and moduli of elasticity of enamel and dentin may be the limiting factors in bond strength. The limitations imposed on the application and performance of a superior adhesive will in effect be a reflection of the inadequacies of tooth structure. Since disease-free teeth survive rather well it must be concluded that these structures are strong enough. Therefore, in speculating on whether an adhesive can do the things we would like it to do the only question is whether an adhesive can match the properties of tooth structure. Adhesives will probably be able to do this, particularly with regard to mechanical properties. Matching coefficients of thermal expansion of tooth structure will probably present a major obstacle. In bacterial resistance and solubility, most resin materials are already superior to enamel and dentin, and if resin materials are the basis of hard tissue adhesives, oral degradation will be unlikely to present a problem. The toxicity potential, particularly from polymerization catalysts and accelerators, is nonexistent with enamel and probably can be made so with vital dentin by careful choice of materials and techniques of application.

There may now exist hundreds of materials that incorporate properties they should have for adhesion to hard tissues. There may also exist the fundamental knowledge necessary to develop a successful dental adhesive. The search for hard tissue adhesives should utilize knowledge gained from available sources and comprise studies of idealized systems and the Edisonian trial and error approaches.

References

Bibby, B. G., and Van Huysen, G. (1933). *J. Am. Dental Assoc.* **20**, 828.
Bowen, R. L. (1961). *In* "Adhesive Restorative Dental Materials" (R. W. Phillips and G. Ryge, eds.), pp. 177-191. Owen Litho Press, Spencer, Indiana.
Bowen, R. L. (1962). U. S. Patent 3,006,112.
Bowen, R. L. (1963). *J. Am. Dental Assoc.* **66**, 57.
Bowen, R. L. (1965a). *J. Dental Res.* **45**, 895.
Bowen, R. L. (1965b). *J. Dental Res.* **45**, 903.
Bowen, R. L. (1967). *J. Am. Dental Assoc.* **74**, 440.
Bradford, E. W. (1967). *In* "Structural and Chemical Organization of Teeth" (A. E. W. Miles, ed.), Vol. 2, pp. 3-32. Academic Press, New York.
Brännström, M. (1960). *Acta Odontol. Scand.* **18**, 17.
Brudevold, F., and Söremark, R. (1967). *In* "Structural and Chemical Organization of Teeth" (A. E. W. Miles, ed.), Vol. 2, pp. 247-273. Academic Press, New York.
Brudevold, F., Steadman, L. T., and Smith, F. A. (1960). *Ann. N. Y. Acad. Sci.* **85**, 110.
Buonocore, M. G. (1955). *J. Dental Res.* **34**, 849.
Buonocore, M. G. (1963). *J. Am. Dental Assoc.* **67**, 382.
Buonocore, M. G. (1964). Unpublished studies.
Buonocore, M. G. (1968). *Intern. Dental J.* **18**, 406.
Buonocore, M. G. (1970). *J. Am. Dental Assoc.* **86**, 324.
Buonocore, M. G., and Bibby, B. G. (1945). *J. Dental Res.* **24**, 103.
Buonocore, M. G., and Casciani, C. (1969). *N.Y. State Dental J.* **35**, 135.
Buonocore, M. G., and Quigley, M. (1958). *J. Am. Dental Assoc.* **57**, 807.
Buonocore, M. G., Wileman, W., and Brudevold, F. (1956). *J. Dental Res.* **35**, 846.
Buonocore, M. G., Azevedo, N., and Dalton, E. (1962). *N.Y. State Dental J.* **28**, 239.
Buonocore, M. G., Matsui, A., and Gwinnett, A. J. (1968). *Arch. Oral Biol.* **13**, 61.
Carderelli, N. F. (1968). "Barnacle Cement as a Dental Restorative Material," Natl. Inst. Health Publ. No. 151. Natl. Inst. Health, Bethesda, Maryland.
Cotton, W. R., Gorman, W. J., and Lamb, J. R. (1965). *J. Dental Res.* **44**, 801.
Cueto, E., and Buonocore, M. G. (1967). *J. Am. Dental Assoc.* **75**, 121.
Eastoe, J. E. (1967). *In* "Structural and Chemical Organization of Teeth" (A. E. W. Miles, ed.), Vol. 2, pp. 279-313. Academic Press, New York.
Frankel, M. A., and Hawes, R. R. (1964). *J. Oral Therap. Pharmacol.* **1**, 147.

Galligan, J. D., Schwartz, A. M., and Minor, F. (1968). *J. Dental Res.* **47**, 629.
Gillery, F. H., and Bush, F. A. (1959). *J. Am. Ceram. Soc.* **42**, 175.
Gillings, B., Buonocore, M., and Sarda, O. (1960). *Dental Progr.* **1**, 61.
Going, R., and Massler, M. (1961). *J. Prosthetic Dentistry* **11**, 298.
Gwinnett, A. J., and Buonocore, M. G. (1965). *Brit. Dental J.* **119**, 77.
Helmecke, J. G. (1967). *In* "Structural and Chemical Organization of Teeth" (A. E. W. Miles, ed.), Vol. 2, pp. 135-162. Academic Press, New York.
Hess, W. C. (1961). *In* "Adhesive Restorative Dental Materials" (R. W. Phillips and G. Ryge, eds.), pp. 10-14. Owen Litho Press, Spencer, Indiana.
Hoekstra, J., and Fritzius, C. P. (1951). *In* "Adhesion and Adhesives" (N. A. De Bruyne and R. Howink, eds.), pp. 33-38. Elsevier, Amsterdam.
Johansen, E. (1967). *In* "Structural and Chemical Organization of Teeth" (A. E. W. Miles, ed.), Vol. 2, pp. 35-72. Academic Press, New York.
Kramer, I. R. H., and McLean, J. W. (1952). *Brit. Dental J.* **93**, 150.
Langeland, K. I. (1957). "Tissue Changes in the Dental Pulp," pp. 74-87. Oslo Univ. Press, Oslo.
Nelson, R. S., Wolcott, R. B., and Paffenbarger, G. C. (1952). *J. Am. Dental Assoc.* **44**, 288.
Newman, G. V. (1965). *Am. J. Orthodontics* **51**, 901.
Pinto, J., and Buonocore, M. (1963). *N.Y. State Dental J.* **29**, 199.
Provenza, D. V., and Sardana, R. C. (1965). *In* "Adhesive Restorative Dental Materials" (H. G. F. Wilsdorf and R. W. Phillips, eds.), Vol. 2, pp. 68-96. U.S. Govt. Printing Office, Washington, D.C.
Quarles, R. W. (1953). *In* "Organic Protective Coatings" (W. von Fischer and E. Bobalek, eds), pp. 167-193. Reinhold, New York.
Relief, D. H., and Dreyer, C. J. (1967). *J. S. African Dental Assoc.* **22**, 338.
Ripa, L. W., Buonocore, M. G., and Cueto, E. (1966). *Proc. Intern. Assoc. Dental Res., Miami Beach, Florida*, p. 100.
Rowles, S. L. (1967). *In* "Structural and Chemical Organization of Teeth" (A. E. W. Miles, ed.), Vol. 2, pp. 201-238. Academic Press, New York.
Scott, D. B. (1960). *Proc. 4th Intern. Congr. Electron Microscopy, Berlin, 1958*, p. 348. Springer, Berlin.
Scott, D. B., and O'Neil, J. (1961). *In* "Adhesive Restorative Dental Materials" (R. W. Phillips and G. Ryge, eds.), pp. 27-38. Owen Litho Press, Spencer, Indiana.
Skinner, E. W., and Phillips, R. W. (1967). "The Science of Dental Materials." Saunders, Philadelphia, Pennsylvania.
Smith, D. C. (1968). *Brit. Dental J.* **125**, 381.
Spreter von Kreudenstein, T. (1953). *Dental Zahnartztl. Z.* **8**, 143.
Swanson, A. T., and Beck, J. F. (1960). *J. Am. Dental Assoc.* **61**, 581.
Swartz, M., and Phillips, R. W. (1960). *J. Dental Res.* **39**, 752.
Takeuchi, M., Kizu, T., Shimizu, T., Eto, M., and Amano, F. (1966). *Bull. Tokyo Dental Coll.* **7**, 60-71.
Trautz, B. R. (1967). *In* "Structural and Chemical Organization of Teeth" (A. E. W. Miles, ed.), Vol. 2, pp. 165-199. Academic Press, New York.
Zettlemoyer, A. C. (1961). *In* "Adhesive Restorative Dental Materials" (R. W. Phillips and G. Ryge, eds.), pp. 88-89. Owen Litho Press, Spencer, Indiana.
Zisman, W. A. (1961). *In* "Adhesive Restorative Dental Materials" (R. W. Phillips and G. Ryge, eds.), pp. 106-136. Owen Litho Press, Spencer, Indiana.

CHAPTER 16

POLYURETHANES

John D. Galligan,
Francis W. Minor,
and
Anthony M. Schwartz

I. Introduction

It has previously been found and reported that certain types of polyurethane resins, formed *in situ*, have the property of adhering with considerable force to freshly cut human dentin or enamel and to maintain this adhesion in the aqueous environment of the mouth (Galligan *et al.*, 1967). These resins are formed by the condensation of a polyglycol with a large molecular excess of a diisocyanate. To take advantage of their unusual adhesive property in developing practical dental restorative systems several basic questions of mechanism must be answered and several operating variables must be optimized. To do this it is necessary to narrow the range of resin compositions being investigated to those having good all around promise and no glaring faults. From among several candidates, we are now reporting on resins having the composition of about 10 molar equivalents of diisocyanate to one molar equivalent of polypropylene glycol of average molecular weight 1000–2000. Since it is not known if this type of urethane resin has the all around properties required of a complete restoration, we have attempted to use it as an adhesive liner for the prepared tooth cavity surface. It thus forms a thin layer between the tooth substance and a conventional restoration material such as amalgam or acrylic polymer. Since this system is intended primarily for anterior restorations emphasis has been placed on the conventional acrylics. For the complete restoration to be effective and durable in the oral environment the liner must adhere to both tooth and acrylic and must in itself have adequate mechanical properties and resistance to the environment. This study accordingly concentrates on tooth–liner interactions and properties of the liner itself.

II. Methods

The methods used to judge the overall effectiveness of a complete restoration (polyurethane liner topped by conventional restorative material) have been described in previous publications (Galligan *et al.*, 1967). They include a tensile test for adhesion, a test for marginal percolation, and tests for durability of the adhesive bond under repeated thermal cycles and compression cycles. In the work covered by this report only the tensile test was used. The procedure for preparing the specimens is as follows:

Freshly extracted sound bicuspid teeth from orthodontic patients were used in all instances. These were mounted in dental stone, and conical cavities were drilled into the teeth to a constant depth, using a round end tapered diamond instrument (Starlite diamond instrument 63, Star Dental Mfg. Co., Philadelphia). The cavities thus cut did not differ significantly in area from tooth to tooth. Since there is no undercutting, and since the conical angle is constant, the force necessary to pull an adhesive restoration out of these cavities is a monotonic (although not necessarily linear) function of the adhesion force.

The liner composition was prepared by thoroughly mixing the isocyanate and polyglycol and waiting until the spontaneously reacting mixture had attained a suitable viscosity. The mixture was then poured or liberally brushed into the cavity and the excess liquid was blown out with an air blast. The amount of liner material left in the cavity formed a layer not less than 10–20 μ thick. After this layer of liner material had been allowed to harden still further (i.e., to become more viscous) for a controlled period of time, the conventional restoration was applied. While the conventional restoration material (acrylic) was still sufficiently soft, the head of a common nickel plated pin was inserted so that the shaft coincided with the conical axis. This "holder pin" provided the means for applying pull on the restoration during the subsequent tensile tests. The restorations were allowed sufficient time to cure and in all instances were stored at least 3 days under water before making the tensile tests on the still wet tooth. Specimens from the same lot were also aged under water for periods up to several months before being tested.

The tensile tests were made on an Instron machine and are reported as total pounds of axial pull necessary to remove the restoration. Because of the geometry of the system, the data cannot easily be translated into numerical values of either tensile or shear resistance of the joint. They should, however, indicate at least semiquantitatively the resistance of the joint to disruption in a practical situation in the mouth. After testing, both the cavity surface and the surface of the restoration were examined to determine the mode of failure. Adhesive failure can occur at the liner-tooth interface or the liner-acrylic interface. Failure within the liner itself is described as cohesive failure. In some specimens two or even all three types can occur, the mode differing from region to region.

In those instances where a preformed restoration was used the procedure was as follows: The conical cavity was prepared in the usual manner, rinsed out with water and dried briefly with an air blast. The acrylic restoration was then formed in the cavity, using the brush technique (Nealon, 1952) and inserting the holder pin as described above. After the acrylic restoration had hardened, it was pulled out of the cavity. The adhesion between the acrylic and the cavity wall was in all instances negligible, and the restorations withdrew easily and cleanly. The cavity was then lined with the liquid urethane composition being tested, the restoration was put back in on top of the liner, and after a suitable hardening and ageing period the tensile test was performed.

Test strips of the urethane compositions themselves, used in examining their mechanical properties, were prepared as follows: The isocyanate-glycol mixture was made and allowed to stand until it had reached a viscosity in the range of 3000–10,000 centipoises. The mixture was then poured onto a clean Teflon plate and spread out to form a layer 1–2 mm thick. This was allowed to stand in air at 92% relative humidity at room temperature for at least 2 days to insure complete cure. The resulting film did not adhere to the Teflon. It was removed and cut into strips of standard width, the thickness of the strips was measured, and they were tested on the Instron machine.

Infrared spectra of the urethane polymers were obtained at various stages of cure as follows: The isocyanate-glycol mixture was prepared and allowed to come to a viscosity of about 3000–10,000 centipoises. This mixture was spread in a thin layer on a rock salt holder. The spectrum was monitored as the curing reaction proceeded at ambient humidity. Alternatively, the mixture was sandwiched between two rock salt holders. In this case, the atmospheric moisture could be taken up only through the edges of the urethane layers and the curing reaction was slower.

III. Materials

The tolylene diisocyanate (TDI) used in this work was Nacconate 80 (Allied Chemical Corp.) a mixture consisting of approximately 80% 2,4-isomer and 20% 2,6-isomer.

PAPI is a polymethylene polyphenylene isocyanate supplied by Upjohn Co., Polymer Chemicals Division, North Hartford, Conn.

Pluracol P-1010 and Pluracol P-2010 are polypropylene glycols of average molecular weight 1000 and 2000, respectively, supplied by Wyandotte Chemical Corp.

A wide variety of other isocyanates and polyols have been used in the course of this work, and are referred to by name where indicated. Qualitative data on these systems have been reported previously (Schwartz *et al.*, 1967).

The acrylic restorations were prepared with Autopolymerizable Liquid (APM methyl methacrylate) and 230 Powder (polymethylmethacrylate containing catalysts) supplied by Monomer-Polymer Division of the Borden Company.

IV. Experimental Procedures and Results

A. Tooth-Liner Adhesions

Effect of Isocyanate-Glycol Ratio

Conventional urethane adhesives are made by the condensation-polymerization of approximately equimolar proportions of glycol and isocyanate. It was discovered relatively early in the present work that a large excess of isocyanate was necessary to attain good tooth-liner adhesion. The effect of varying the isocyanate-glycol ratio in three different systems is shown in Table I. In both the TDI-P-2010 and the TDI-P-1010 systems the primary isocyanate-glycol mixtures were allowed to come to 3000-6000 centipoise viscosity before being applied to the freshly cut cavity surface. This required 15-40 minutes, depending on the ambient humidity. The acrylic restoration was prepared on top of the liner immediately after the latter had been applied to the tooth. The completed assembly was aged overnight at 92% humidity at room temperature to insure complete hardening of both the liner and the acrylic. The assembly was then stored under water at 40°F for 3-7 days (within this interval the time of storage had no significant effect) before being tested. The PAPI-P-1010 systems were prepared and tested in the same way, except that the primary mixture was allowed to come to a viscosity of 4000-10,000 centipoise, which required 1-3 hours, before being applied to the cavity surface.

The data in Table I show that in all three systems it is necessary to have a large excess of isocyanate in the system to develop satisfactory adhesion. However, if the excess goes too high (in these cases to 20 moles isocyanate per mole glycol), the adhesion again diminishes.

TABLE I
THE EFFECT OF EXCESS ISOCYANATE ON THE ADHESION OF POLYPROPYLENE GLYCOL URETHANES TO TEETH[a]

NCO/OH (mole/mole)	Polyurethane	Tensile strength (lb)
1	TDI-P-2010	0
2	TDI-P-2010	0
5	TDI-P-2010	7.5
10	TDI-P-2010	8.3
15	TDI-P-2010	7.1
20	TDI-P-2010	0.2
1	TDI-P-1010	0
5	TDI-P-1010	1.7
10	TDI-P-1010	6.2
20	TDI-P-1010	0.3
1	PAPI-P-1010	0
2	PAPI-P-1010	0
10	PAPI-P-1010	6.5
20	PAPI-P-1010	1.2

[a]Each polyurethane was used as a liner for acrylic restoration and was tested after 3-7 days storage in water.

B. Effect of Preapplication Viscosity

It was expected that the adhesion of liner to tooth might be affected by the extent to which the isocyanate-glycol reaction had taken place before the mixture was applied to the tooth. Adhesion might also be affected by the smoothness of the prepared cavity surface. To test these possibilities a set of cavities prepared as above was divided into two groups. One was left in the normal state, designated "rough." The other was polished to as smooth a finish as possible using a wet felt impregnated with γ-alumina. Liners were made up from 10 molecular proportions of isocyanate and 1 of glycol, and were allowed to age for differing periods. In this way, different viscosities were attained, before the liners were applied. The acrylic restorations were then applied and cured in the manner described above; the assembly was stored in water 4–7 days and then tested.

The results are shown in Table II. In the TDI systems there is little

TABLE II

THE EFFECT OF PREPOLYMER VISCOSITY AND CAVITY ROUGHNESS ON THE ADHESION OF POLYURETHANE LINERS TO TEETH

Polyurethane	Viscosity (centipoise)	Cavity surface	Tensile strength (lb)
TDI-P-2010	130	Rough	14.8
TDI-P-2010	2980	Rough	12.7
TDI-P-2010	5730	Rough	10.3
TDI-P-2010	10,900	Rough	0.7
TDI-P-2010	150	Polished	4.3
TDI-P-2010	3210	Polished	7.2
TDI-P-2010	5900	Polished	8.5
TDI-P-2010	11,600	Polished	8.8
TDI-P-1010	98	Rough	16.0
TDI-P-1010	290	Rough	10.2
TDI-P-1010	2100	Rough	8.6
TDI-P-1010	8150	Rough	6.8
TDI-P-1010	20	Polished	3.7
TDI-P-1010	560	Polished	8.8
TDI-P-1010	4240	Polished	7.8
TDI-P-1010	10,000	Polished	10.6
PAPI-P-1010	250	Polished	5.3
PAPI-P-1010	6540	Polished	4.6
PAPI-P-1010	10,000	Polished	5.8
PAPI-P-1010	154,000	Polished	1.7

or no significant effect of preapplication viscosity on the final adhesion until the viscosity has reached a value over 10,000 centipoise. It is noteworthy that these TDI-glycol mixes begin to get opaque and spongy and partly solidified when the measured viscosity goes much over 10,000 centipoise. If they have reached this stage before being applied, the adhesion is usually poor and tensile failure occurs at the liner-tooth interface. Otherwise the failure is generally cohesive within the liner. The PAPI-glycol mixtures show no such phase separation as they age, but become progressively more viscous while remaining clear and homogeneous up to the region of solidification. The data show that viscosity has little effect on adhesion except when the mixture has become so viscous (154,000 centipoise) that it is difficult to apply evenly and thoroughly to the cavity surface.

C. Effect of Surface Roughness

The data for both TDI systems (Table II) indicate that the rough surfaces develop higher adhesion than the smooth ones, particularly when the isocyanate-glycol mixture is applied at a low viscosity. This is to be expected. Rough surfaces have a larger actual area per unit apparent area than smooth ones and, therefore, have greater potential adhesion. A low viscosity adhesive can flow into the rugosities and thereby take advantage of the extra available area of the rough surface.

D. Effect of Moisture of the Tooth-Liner Interface

Tooth substance is generally considered hydrophilic, and in the presence of water we might expect difficulty in attaching nonaqueous polymers such as these urethanes to the cavity surfaces. The isocyanate radical itself reacts with water, and we might, therefore, expect the course of the condensation-polymerization reaction between isocyanate and glycol to be modified (possibly adversely) by the presence of water. The effect of water at the cavity-liner interface during application of the liner was studied in a TDI-P-2010 system and a PAPI-P-1010 system. Cavities were prepared in the usual manner, rinsed free of debris with water, and dried as shown in Table III before applying the isocyanate-glycol mixture. The undried specimens had liquid water showing on the cavity surface when the liner mixture was applied. The specimens dried in a desiccator over calcium chloride had lost over 8% of their original weight and could be considered substantially "bone dry" at the time the liner was applied. The air blasting removed liquid surface water only, and the dry nitrogen blast removed some of the interior water. Both liner mixtures had an isocyanate to glycol molar ratio of 10:1. At the time of application the TDI sys-

TABLE III
THE EFFECT OF MOISTURE ON THE STRENGTH OF POLYURETHANE-LINED ACRYLIC RESTORATIONS

Cavity preparation	Weight loss (%)	Polyurethane	Relative humidity (%)	Tensile strength (lb)
Not dried	0	PAPI-P-1010	92	4.7
2 Sec air blast	0	PAPI-P-1010	92	7.1
5 Min N_2 blast	0.5	PAPI-P-1010	92	5.2
8 Days over $CaCl_2$	8.3	PAPI-P-1010	92	7.4
8 Days over $CaCl_2$	8.3	PAPI-P-1010	0	0
Not dried	0	TDI-P-2010	92	5.7
2 Sec air blast	0	TDI-P-2010	92	4.8
5 Min N_2 blast	0.5	TDI-P-2010	92	5.9
8 Days over $CaCl_2$	8.3	TDI-P-2010	92	7.4
8 Days over $CaCl_2$	8.3	TDI-P-2010	0	0

tem had a viscosity of 3000-6000 centipoise and the PAPI system, 4000-8000 centipose. After acrylic restorations were applied on top of the liners, all specimens except one of the bone dry ones from each system were allowed to cure 24 hours at 92% relative humidity, and were stored 4 days under water before tensile testing. The two bone dry specimens were stored 5 days at zero humidity after their liners and acrylic restorations had been put in place, and they were then subjected to the tensile test without ever having been exposed to moisture vapor or water.

The results of the tensile tests are shown in Table III. The liners in the water-free specimens never did cure and were still liquid at the time of testing. The tensile strength of the joints were essentially zero. Among the other specimens there was no significant difference in tensile strength. Those in which the liner had been applied to the wet cavity surface were just as strong (statistically) as those which had been dried to varying degrees beforehand. Thus, the presence of water does not seem to affect the adhesion adversely. In the total absence of water, moreover, the urethane polymer is not formed at any reasonable rate and adhesion in the joint is not developed at all.

V. Properties of the Liner

A. Mechanical Property Studies

Strips of liner material approximately 1 mm thick were prepared as described in the methods section above. The strips were stored under water, and at intervals of 1-130 days specimens were taken for testing.

All urethanes in this study contained 10 moles isocyanate to 1 mole glycol. Data for the PAPI-P-1010, PAPI-P-2010, and TDI-P-2010 systems are shown in Tables IV, V, and VI, respectively. First, all the data show that the tensile properties of the final cured polymers are not affected significantly by long immersion in water. Second, the PAPI systems have much higher tensile strength than the TDI systems. Within the PAPI systems, the P-1010 combination has higher strength

TABLE IV
WATER RESISTANCE OF PAPI-PLURACOL P-1010 FILMS[a]

Film composition (NCO/OH)	Age (days)	Tensile strength (lb/in^2)	Elongation (%)
10	4	7300	7.0
10	7	7590	7.5
10	14	6700	4.0
10	29	8140	6.9
10	74	8450	8.0
10	105	7700	5.9
10	119	6800	4.9
4	7	1210	50
4	14	2000	60
4	29	2330	52
4	74	2130	49
4	119	2180	42

[a]Each film contained PAPI and Pluracol P-1010 as indicated and was stored in distilled water at 25°C for the period indicated.

TABLE V
THE WATER RESISTANCE
OF PAPI-PLURACOL P-2010 FILMS[a]

Age (days)	Tensile strength (lb/in^2)	Ultimate elongation (%)
1	2000	21
4	2540	12
7	2260	28
10	2980	11
14	2140	17
17	3660	23
25	2940	12
29	2020	16
51	3220	15
73	2300	22
120	2900	18

[a]Each film contained 3 parts PAPI and 2 parts Pluracol P-2010 (10:1 molar) stored at 25°C in distilled water for the period indicated.

TABLE VI
WATER RESISTANCE OF
NACCONATE 80-PLURACOL P-2010 FILMS[a]

Age (days)	Tensile strength (lb/in^2)
5	460
17	361
37	428
75	496
112	445

[a]Each film contained equal parts Nacconate 80 and Pluracol P-2010 and was stored in distilled water at 25°C for the period indicated.

and lower elongation than the P-2010 combination at the same molar ratio. Increasing the proportion of glycol to isocyanate lowers the tensile strength but increases the elongation.

The urethane systems of this study require water for rapid and complete cure, and a byproduct of the reaction with water is carbon dioxide. The liberation of carbon dioxide can cause bubbles and flaws in the final resin. The possibility was studied of eliminating the carbon dioxide by incorporating calcium hydroxide in the mixture. A set of strips was made using the TDI-P-1010 system and incorporating enough $Ca(OH)_2$ to react with all the CO_2 that might be given off. Those strips were tested after progressive intervals of ageing in water. The results are shown in Table VII. The strips which had no $Ca(OH)_2$ showed no change in tensile strength on prolonged exposure to water, confirming previous observations. The strips containing $Ca(OH)_2$ were significantly stronger than the unlined strips just after preparation. Prolonged exposure to water weakened them greatly, however. This is to be expected since the urethane and urea linkages are known to be susceptible to hydrolysis by aqueous alkali.

B. Spectroscopic Studies

A 10:1 molar mixture of TDI and P-2010 was spread on a rock salt plate and allowed to cure at ambient humidity and temperature, as described in the methods section. Three infrared bands were monitored: 3285 cm^{-1}, assigned to NH, which is present in both urethane and substituted urea groups; 1710 cm^{-1}, characteristic of urethane carbonyl; and 1640 cm^{-1}, assigned to substituted urea carbonyl (Szymanski, 1964). The bands due to isocyanate groups were also monitored. The results are shown in Fig. 1. The urethane content of the mixture increased rapidly during the first two or three hours, presumably until all the hydroxyl groups are used up, then remained constant. The NH and substituted urea carbonyl contents continued to increase with time. The isocyanate content (not shown in Fig. 1) decreased with time, but some isocyanate was still present after two weeks. Eventually all isocyanate disappeared. In the sandwich system (isocyanate-glycol between two rock salt plates) the reactions were essentially similar, but slower.

VI. Discussion

It is well-known that many chemical changes occur when a diisocyanate reacts with a glycol, and that the reactions are even more complicated when water is involved. It is evident that water is essential in forming the adherent liner systems described above, at least if they are to be formed at room temperature in a reasonable time period. Of the many molecular species that might be formed, it appears that one or more have the desired property of being adherent to tooth substance in the presence of water, so much so that a strong adhesive bond is formed even on a wet tooth surface. It is of obvious importance to discover what species these are. Once their identity is established, efforts can be made to increase their concentration in the final

TABLE VII

THE EFFECT OF CALCIUM HYDROXIDE ON THE STRENGTH ON NACCONATE 80-PLURACOL P-1010 FILMS[a]

Age (days)	$Ca(OH)^2$(%)	Tensile strength (lb/in^2)
3	0	680
	26.8	1040
12	0	655
	26.8	980
33	0	615
	26.8	680
59	0	725
	26.8	470
130	0	680
	26.8	370

[a]Each film contained 2 parts Nacconate 80, 1 part Pluracol P-1010 (10:1 molar), and calcium hydroxide as indicated. Each film was stored in distilled water at 25°C for the period indicated.

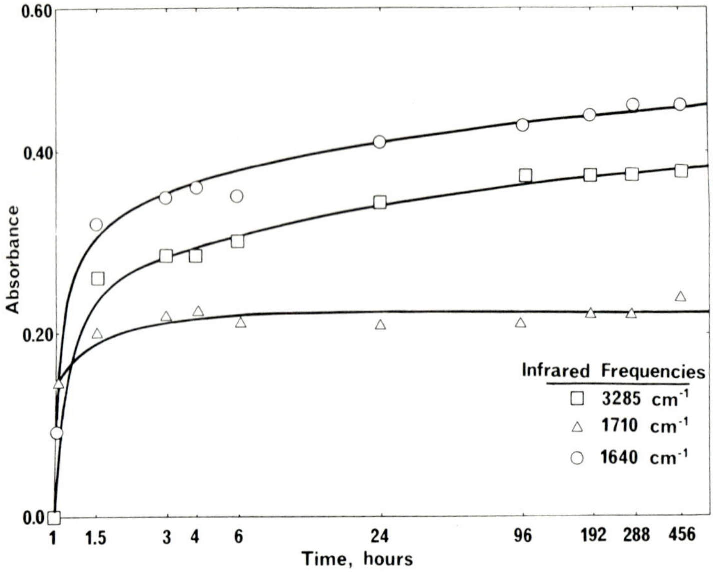

Fig. 1. Results of spectroscopic studies. (See text.)

cured liner, and if necessary, to increase their long-term stability in the oral environment.

Given a 10:1 molar mixture of a diisocyanate and diglycol in the presence of water, the following known reactions can occur:

$$n\,\mathrm{OCN - R_1 - NCO} + n\,\mathrm{HO - R_2 - OH} \rightarrow (-\overset{\mathrm{O}}{\overset{\|}{\mathrm{C}}}\mathrm{H} - \mathrm{R_1} - \mathrm{NCOR_2} - \mathrm{O})_n \quad (1)$$

(I)

$$\mathrm{OCN - R_1 - NCO + H_2O \rightarrow OCN - R_1 - \overset{H}{N}COOH} \quad (2)$$

(II)

$$\mathrm{II \rightarrow CO_2 + OCN - R_1 - NH_2} \quad (3)$$

(III)

$$n(\mathrm{III}) \rightarrow (\overset{\mathrm{O}}{\overset{\|}{\mathrm{C}}} - \underset{\mathrm{H}}{\mathrm{N}} - \mathrm{R_1} - \underset{\mathrm{H}}{\mathrm{N}} -)_n \quad (4)$$

(IV)

Starting with 10 molar equivalents of isocyanate groups and 1 molar equivalent of OH groups, the primary product is the urethane (I) formed according to reaction (1). This leaves 9 equivalents of isocyanate groups to react with water (ambient or deliberately added) according to reaction (2). The carbamic acid (II) decarboxylates according to reaction (3) to produce amino groups, which can in turn react with isocyanate groups according to reaction (4) or its equivalent. If reactions (3) and (4) are faster than reaction (2), only 4.5 moles of isocyanate groups will react with water, the remainder reacting with the NH_2 groups. Meanwhile, 4.5 moles of CO_2 would be liberated. Preliminary studies have indicated that somewhat less than this amount of free CO_2 is actually produced during the cure of the liner material. These reactions are currently being studied in greater detail, in an effort to identify the tooth-adherent species and clarify their chemistry.

Acknowledgments

This investigation was conducted under USPHS Contract No. PH-43-64-530 entitled "An Investigation of the Mechanism of Adhesion to Teeth" from the National Institute of Dental Research, National Institutes of Health, Bethesda, Maryland.

References

Galligan, J. D., Schwartz, A. M., Minor, F. W., and Nelsen, R. J. (1967). *Abstr. 45th Gen. Meeting Intern. Assoc. Dental Res., 1967, Washington, D.C.* p. 56. Univ. Micro-Films, Ann Arbor, Michigan.

Nealon, F. H. (1952). *J. Prosthetic Dentistry* **2**, 513–527.

Schwartz, A. M., Galligan, J. D., and Minor, F. W. (1967). *Natl. Inst. Dental Res. Publ.* **PV 174848.**

Szymanski, H. A. (1964). "Interpreted Infrared Spectra." Plenum Press, New York.

CHAPTER 17

SURFACE PREPARATION AND VARIOUS ADHESIVE RESINS

Henry L. Lee, Jr.
and
Michael L. Swartz

I. Introduction

The medical and dental technologies would vastly benefit from polymers capable of bonding to hard tissue. The orthopedic field, for example, has an urgent requirement for long-term adhesives for bone repair and positive attachment of prosthetic devices. Arthroplasty applications include reconstruction of the femur head or fractured humerus and attachment of artificial ligaments or synthetic tendons to bone. The neurosurgeon performing a craniotomy would prefer adhesion to the cranium for filling trephinement cavities or attachment of larger prosthetic repairs.

Practitioners of preventative dentistry need a low viscosity adhesive for filling developmental pits and fissures. The orthodontist requires a temporary adhesive to replace the bulky, unhygienic, and unsightly metal bands now employed to attach brackets to the enamel. Periodontists could use a dental adhesive for splinting teeth (the positional fixation of teeth during healing of soft tissue and alveoli), and prosthodontists could use one for improved long-term cementation of metal restorations. An adhesive dental filling material is in great demand to provide long-term caries free restorations.

Types of cavity preparation presently employed are by necessity designed to do more than simply remove decayed tooth structure. It is often necessary to remove healthy tooth either to eliminate danger of fracture of weakened or undermined tooth structure, extend cavity walls into areas where margins may be kept clean, or provide for mechanical retention of a nonadhering filling material. In addition, adhesion of the filling material to the surrounding cavity walls would reduce or eliminate migration of oral fluids and bacteria into the restorative tooth interface and thus reduce the incidence of recurrent caries.

Modern theories of adhesion may be summarized by the following statement: *If a liquid can be caused to wet a substrate and then caused to convert from the liquid phase to the solid phase with low enough shrinkage so that the adsorption forces between the liquid and substrate are not broken or disrupted, then the solidified liquid will bond to the substrate.* Although this is a simple sounding rule of thumb, it is full of pitfalls. The definition of the term "wetting," for example, involves the complex science of thermodynamics of surfaces, involving surface tension, critical surface tension, surface free energy, contact angle (dynamic and static, advancing and receding), cohesive energy density, spreading pressure, solubility, and gas adsorption.

Further, the problem of predicting and measuring shrinkage is also complex. There is no theory to predict adequately what molecular configuration will reveal minimum shrinkage during polymerization or freezing. Methods of measuring shrinkage are either best suited to large masses or are time consuming and expensive to conduct when attempting to measure the properties of the thin films encountered in glue lines. In the case of restorative dental materials, the volumetric skrinkage of the filling material (varying from 0.4 to 8%) (Smith and Schoonover, 1953; Bowen, 1963) can produce tensile stress at the interface of the tooth and the restorative material if that material is adhesive, or it will pull away from the tooth (Fig. 1) if it is not. Adequate

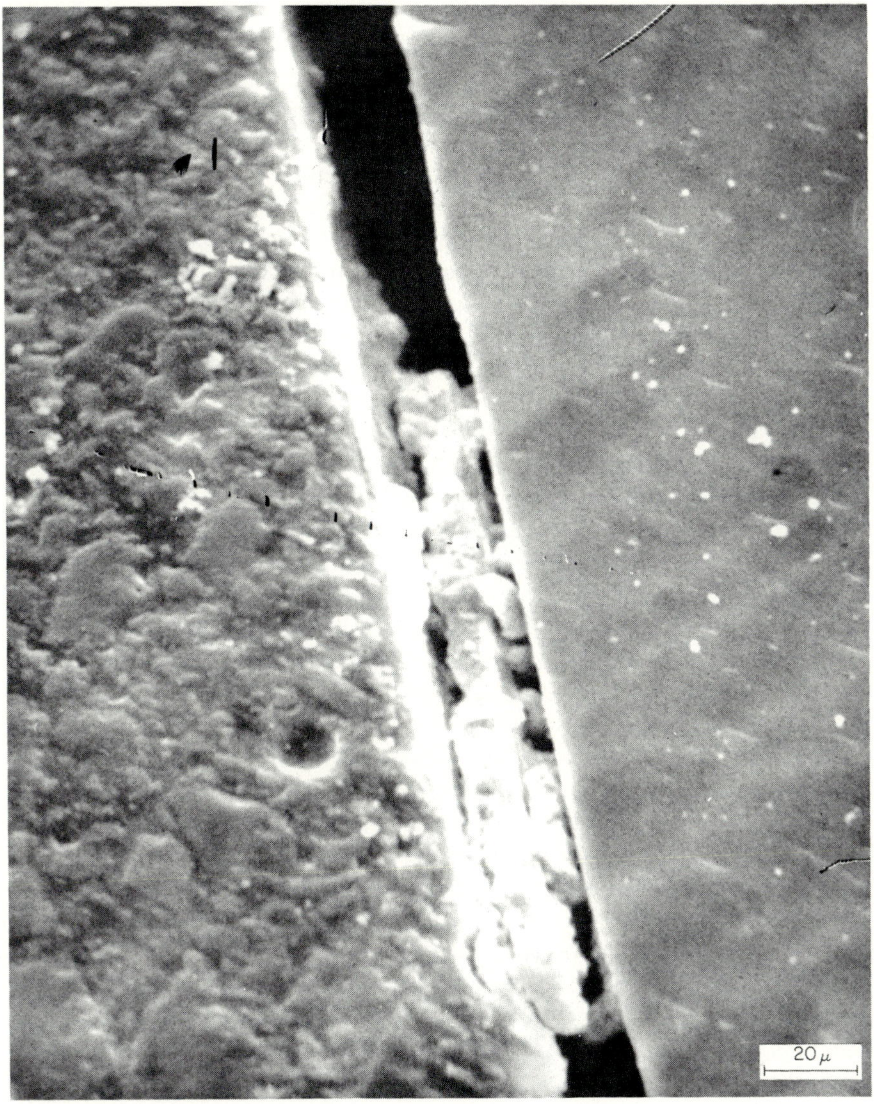

Fig. 1. Scanning electron photomicrograph of the interface between a low shrinkage (0.5%) dental restorative material and the cavity base. ×300.

adhesion must compensate for this tensile stress if the cavity is to remain sealed. Bowen (1967) found that shrinkage can exert forces ranging from 250 to over 1000 lb/square inch depending upon the size of the cavity and the volumetric shrinkage of the material.

Hence, industrially, an empirical, cut-and-try approach is used. Potential materials are tested as adhesives, and the better formulations are gradually adopted for use. When one attempts to adopt the industrial approach for hard tissue, either teeth or bones, further complexities are encountered in trying to use the rule of thumb on adhesion. There are two unstated assumptions, namely, that (1) the surface to be bonded is clean of debris, such as greases, oils, and other nonadhesive substances, so that a uniform surface exists for bonding; and (2) the surface must be representative of the bulk, cohesive solid material. The latter point is important in order that the surface be strong and dimensionally stable enough to constitute a suitable adhesive substrate for the structural loads which the adhesive joint will transmit to the substrate. Unfortunately, from an adhesives chemist's point of view, hard tissue is a rather poor substrate. It is a complex, composite, hydrophilic material, containing fats, lipids, collagen, and water soluble crystals. The surface of a prepared bonding area is also usually smeared with debris from the sawing or grinding operation (Fig. 2). Both of these situations tend to violate the unwritten assumptions that the surface be clean and that the substrate be solid and cohesive.

The biggest problem undoubtedly is the hydrophilic nature of tissue. Living systems are essentially aqueous colloids. In the tooth, water is fed by the tubules right up to the glue line; and water has a higher energy of adsorption to most crystals than do polymers, so that water tends to displace adsorbed adhesive polymer molecules. Nature has overcome this problem rather well by the use of water soluble polymer intermediates and polyelectrolytes as polymers, which serve to set up strong adhesive forces even in wet fields or colloidal situations. However, when these substances are synthesized and deposited in the very unsophisticated manner available today, they do not yield the high bond strengths achieved in nature. Probably techniques such as charge or electrical field controlled polymerization of polyelectrolytes will in time make it possible to grow or deposit polymers *in situ* in a manner which will give markedly enhanced bond strengths. But even these heroic measures of application may not be adequate.

In many cases, the polymer is asked to serve as a gap filler or replacement for bone or dentin or enamel, and the synthetic polymer is required to possess mechanical properties superior to those of both the natural polymer and the polymer composite. The problem is difficult.

This discussion of bonding to hard tissue will be primarily concerned with teeth. The problems encountered in bonding to bone are

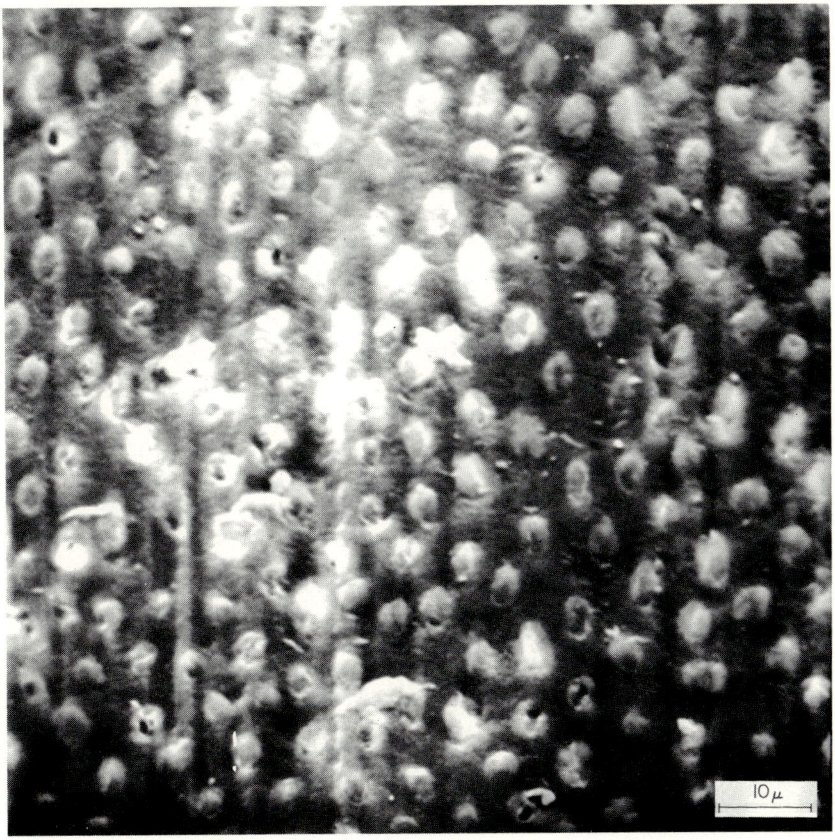

Fig. 2. Scanning electron photomicrograph of ground dentin surface. Note the tubules filled with grinding debris. ×1000.

essentially the same as those with teeth. Unlike teeth, however, bone tends to regenerate itself after it is broken, and its ideal adhesive would consequently be one that would not interfere with the natural regeneration, and as such would be replaced progressively by ingrowing bone. In adhesive systems which have been investigated in the past, the adhesive tends to be encapsulated which forms a barrier for subsequent ingrowth and calcification. Attempts to make the prosthesis porous to facilitate ingrowth have not been entirely satisfactory and, in the absence of ingrowth, adequate strength has not been obtained.

Craniotomy plates, even when cured essentially *in situ*, are often removed after cure, drilled, and then sutured into place with stainless steel sutures. When acrylic pastes are used to hold femoral head prostheses in the intermedullary canal, the cured paste achieves its function by mechanical actions and is usually encapsulated and shielded from direct contact with the bone by this fibrotic capsule. Bone is also vital tissue which bleeds when cut or broken, and materials are needed which simulataneously provide hemastasis.

II. The Substrate

Because it is possible to manipulate or to control the properties of hydrophobic polymers quite well, a logical approach to obtaining adhesion to hard tissue is to develop means for surface preparation and reinforcement of the substrate. In other words, what can be done to the surface to make it more suitable for bonding, and more suitable for long-term bond retention, so that the best use can be made of the adhesive, high strength polymers such as epoxy and polyurethane resins available today?

In most dental applications the tooth is ground or cut before the bonding operation. This operation leaves in its path smeared, loose and denatured tooth debris (Provenza and Sardana, 1965), which must be removed if optimum adhesion is to be maintained. An example of the result from bonding to a ground dentin surface which was not cleaned of debris is shown in Fig. 3. This scanning electron photomicrograph shows that the adhesive had bonded quite well to the substrate, but approximately 50% of that substrate was loose debris from the grinding operation. The tensile strength of this adhesive failure was in fact about one-half the bond strength of the same adhesive when applied to a cleansed dentin surface.

A. *Removing Debris*

Many investigators have studied the problem of removing debris. The principal methods used have involved a protocol such as:

(1) Scrubbing with detergent
(2) Washing with hydrogen peroxide
(3) Etching with dilute hydrochloric acid
(4) Treating with ethylenediaminetetraacetic acid
(5) Sweeping with an antistatic brush

Other cleansing protocols have involved topical applications of reagents such as: (1) dilute hydrochloric acid, and (2) 50% H_3PO_4.

In the authors' laboratory, adhesive strength values have been used as a means of evaluating the efficiency of various cleansing agents.

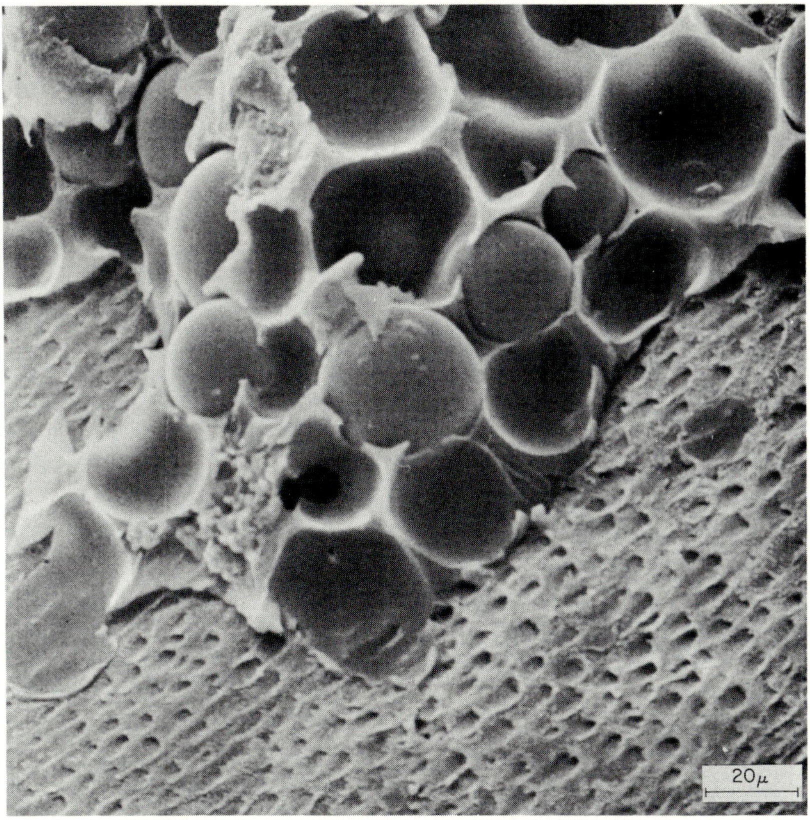

Fig. 3. Scanning electron photomicrograph of a partially cohesive tensile failure.

The effectiveness of each step of cleansing is usually checked under a microscope, and also by means of contact angle of a drop of uncatalyzed resin on the surface. Reagents intended to attack the collagen debris and the hydroxyapatite debris have been evaluated. Enzymes were selected to attack collagen because they are appreciably faster than most laboratory reagents normally used to hydrolyze amide or peptide linkages. To attack hydroxyapatite debris, various chelating agents were studied: EDTA, citric acid, oxalic acid, and others. These ingredients have all been suggested as adhesion improving primers or preconditioners in the literature (Chang and Uy, 1965), but have not been cited as debris removers.

A solution of trypsin was found especially effective in removing debris, as shown in Fig. 4. Note especially the cleansing of the tub-

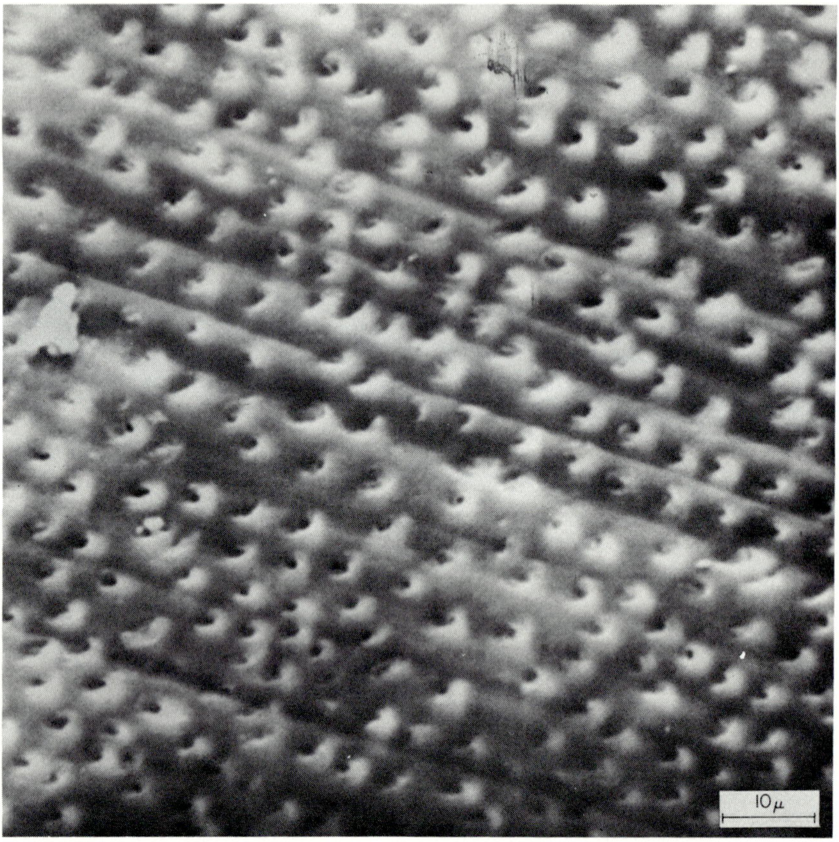

Fig. 4. Scanning electron photomicrograph of the effects of trypsin on dentin. ×1000.

ules. Of the reagents investigated for removing hydroxyapatite debris, citric acid and phosphoric acid were found also to be effective (Fig. 5).

The general performance of various reagents may be summarized as follows:

Cleansing Action of Reagents on Hard Tooth Structure under Clinical Time Limitations

Good	Fair	Poor
Proteinases	Acetic acid	Oxalic acid
Phosphoric acid	EDTA	Hydrogen peroxide
Citric acid	Boric acid	Sodium hydroxide
Sulfuric acid	—	Hydrochloric acid

B. Cross-linking of Collagen

The principal material making up the organic matrix of hard tissue is the protein collagen. In dentin it is readily permeated by various reagents. Figures 6 and 7 show the penetration of two dye solutions into the surface of dentin to a distance of 0.3 mm. These photomicrographs suggest the use of an impregnating solution capable of cross-linking the collagen and rendering it hydrophobic. There are numerous reagents which will do this, but most of them require prolonged periods of application, are very toxic, or do not form stable bonds. Thus, dialdehydes cross-link collagen readily, but the resulting Schiff

Fig. 5. Scanning electron photomicrograph of the cleansing action of citric acid upon dentin. ×1000.

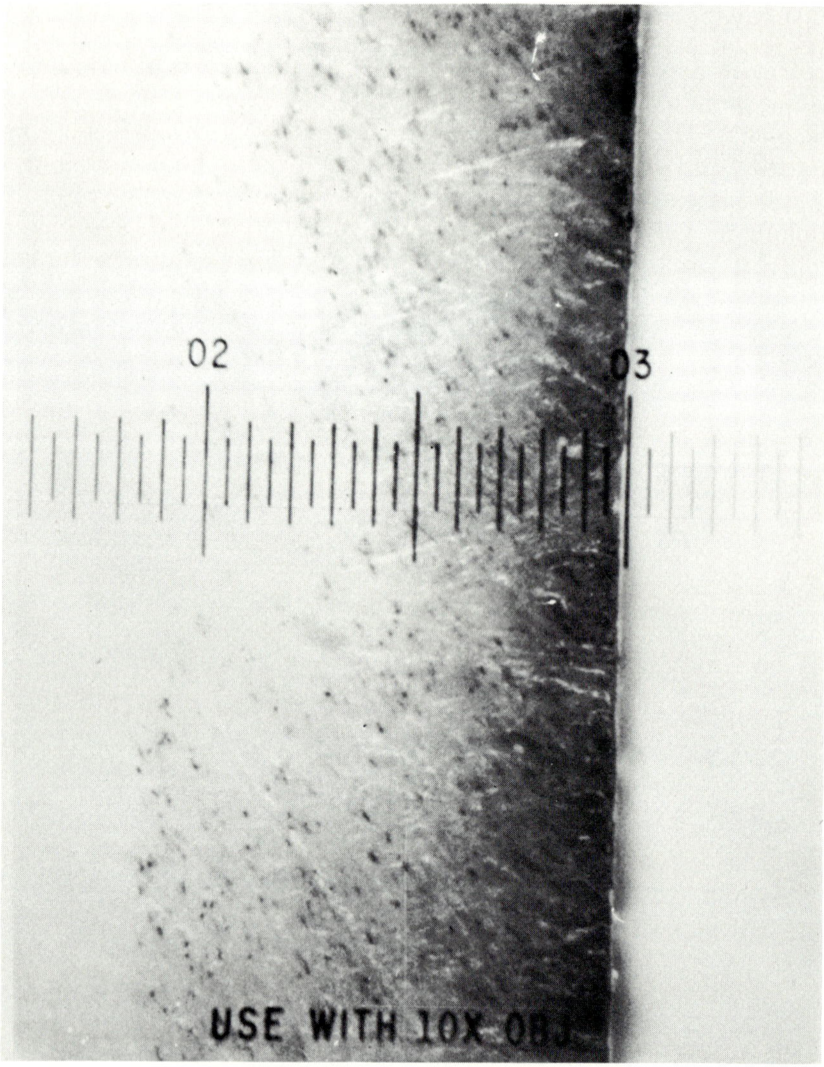

Fig. 6. Scanning electron photomicrograph of sectioned tooth structure showing the penetration of 2-methyl-2-butanol. ×170.

bases are readily hydrolyzed by water. Diisocyanates and zirconium salts are possibilities. Silver nitrate has also been suggested.

Since the protein collagen is 18% by weight and 39% by volume of the dentin, a large portion of the surface area at the bonding surface is an organic matrix which is porous and thus permeable to water. If the only adhesion is that between the adhesive material and the hydroxyapatite of the dentin, then strength and permanency of the adhesive bond are markedly reduced. Reduction results, first, because only one-half to two-thirds of the surface area is available for bonding, and

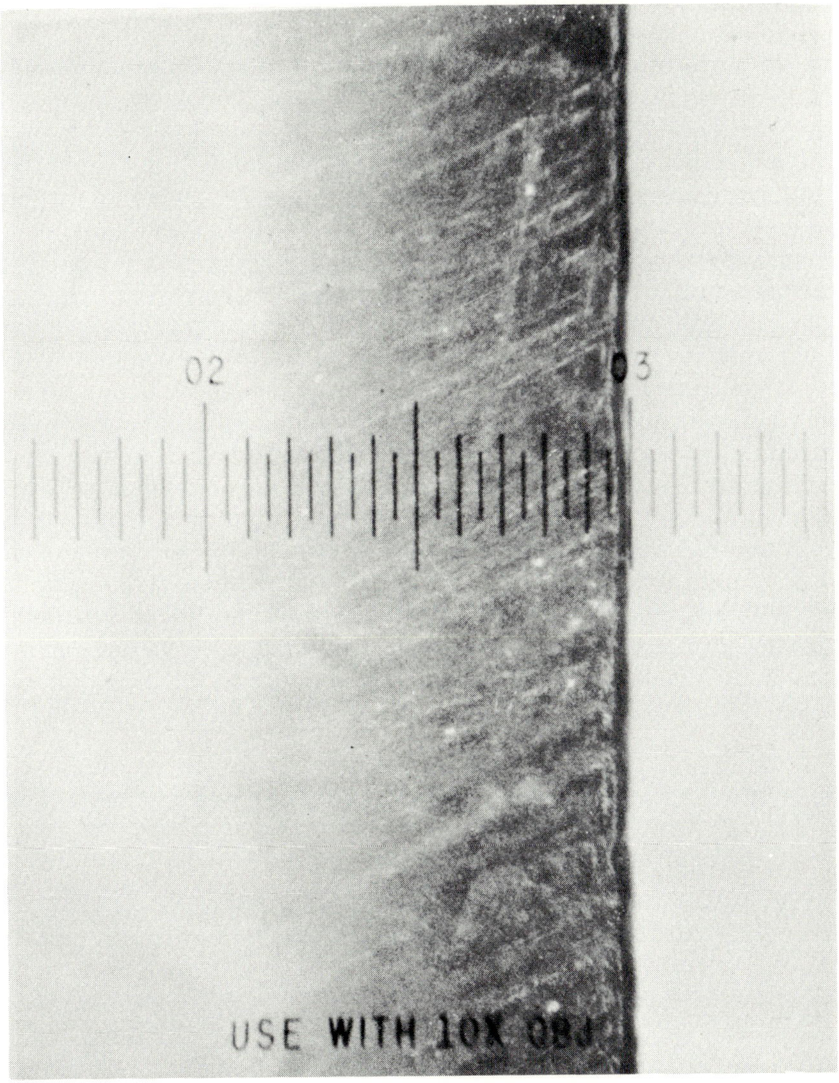

Fig. 7. Scanning electron photomicrograph of sectioned tooth structure showing the penetration of silane. ×170.

second, because the bond between the resin and each crystal of hydroxyapatite is surrounded by a matrix which is, in effect, feeding water directly to the sides of each minute bond interface. This can disrupt the adhesive bond and weaken the collagen. Studies in our laboratories with reconstituted collagen, for example, have shown that whereas dry collagen has a tensile strength of 7500 lb/square inch, wet collagen has a tensile strength of only 100 lb/square inch.

In order to achieve maximum adhesion to the tooth structure and maximum strength in the bond, it would be desirable to develop methods of rendering the surface of the collagen bondable by improving and sealing the collagen layers so that water will not permeate to the interface. Because of the readily available reactive sites in collagen (amine-hydrogens, amide-hydrogens, hydroxy groups, and mercaptan groups), it should be possible to cross-link the collagen or to render it hydrophobic.

One class of cross-linking agents, recognized as being very reactive, contains zirconium salts. Its distinguishing features in compounds are its polyvalence; formation of stable and durable bonds with electronegative elements—particularly those of the second period (oxygen, nitrogen, and fluorine); extreme stability of the 4-oxidation state; and extreme resistance to bonding with carbon. The compounds are also nontoxic and colorless. From this, it might be anticipated that materials containing oxygen, nitrogen, or fluorine atoms can be cross-linked by zirconium atoms in two or three dimensions, without side reactions and without imparting color or toxicity to the cross-linked substrate. Linkages of zirconium with oxygen-containing organic groups can be of the following types:

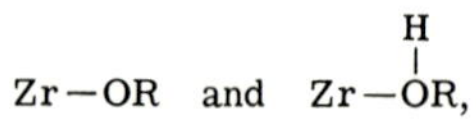

and with other zirconium atoms of the following types:

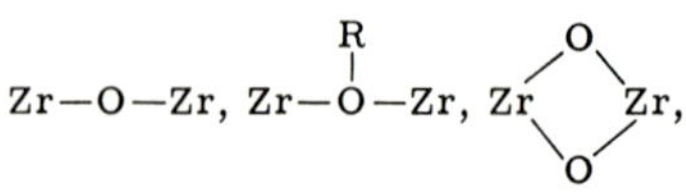

and

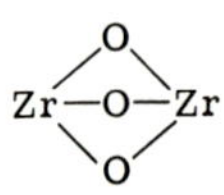

Which type will occur and predominate depends on the geometry of the molecules with which the zirconium atoms combine and on environmental conditions such as pH and the presence of various ions which affect the charge on zirconium, oxygen, nitrogen, and certain other atoms which may become involved in the linkage. Oxygen-containing anions such as acetate, sulfate, and carbonate, complex with the oxozirconium cation and neutralize its electrical charge or change the charge from positive to negative, depending upon the proportion of anions available for complexing.

Sulfatozirconate and acetozirconate ions formed in this fashion are particularly reactive with protein molecules. They have been used to harden gelatin in a reaction analogous to the proposed method of hardening or cross-linking tooth collagen. The most likely compound for this study has been the carbonatozirconate or ammonium zirconyl carbonate which has the formula:

$$(NH_4)_3ZrOH(CO_3)_3 \cdot 2H_2O$$

It was found that the zirconium tetraacetate at pH 4.7 causes immediate gelation of powdered collagen. This gelling may be indicative of cross-linking caused by zirconium since addition of a similar solution without zirconium produced no effect.

C. Priming Hydroxyapatite

Industrially, in the fabrication of epoxy resins reinforced with glass fiber, it has been found that chromium complex and silane treatments of the fibers not only increase adhesion, but also cause retention of strength when the specimens are exposed to water. Thus, whereas laminates without silane treatment retain their strength for only 1 year in hot water, silane treated laminates gave strength retention estimated to be effective for 50 years, based on accelerated tests.

Tests of chromium complexes and a number of organofunctional silanes revealed that several of the silanes did enhance adhesion to tooth structure. The postulated mechanism is shown in Fig. 8. Typical results are shown in Fig. 9.

D. Drying of the Surface

A vital tooth contains over 20% water. Bonding to such a wet substrate presents a problem. Most epoxy resins which have the other basic properties for dental adhesion, will not adhere to enamel or dentin unless the surface is dried first. Drying with heat is not practical, and most organic solvents which have been used to degrease and dry do not effectively displace water for a sufficient length of time to be

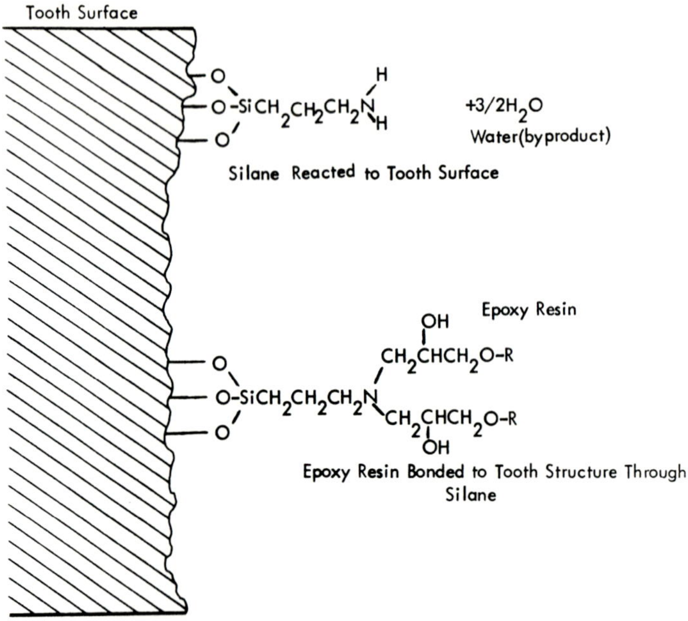

Fig. 8. Reaction between silane coupling agent, tooth structure, and epoxy resins.

practical. Zisman (1965), however, showed that an effective water displacing agent should have (a) a spreading pressure on water above 25 dynes/cm, (b) a boiling point above 90°C, and (c) a solubility in water between 2 and 25%. Several aliphatic alcohols were found, in this respect, to be effective. One of the alcohols, 2-methyl-2-butanol, was tested on teeth and found to displace water in dentin to a depth of 0.3 mm for as much as five minutes.

E. Test Specimens

Testing of adhesives for living systems requires extensive *in vitro* testing which assumes a correlation to *in vivo* or clinical testing. Ideally, human teeth should be used as specimens, and one may collect these from neighboring dental surgeons. Unfortunately, they vary widely in age, size, and physical condition due to the presence of restorations, carious lesions, etc., so that dozens of teeth must be processed to obtain a good specimen. Suitable test specimens from animals can pose similar problems. However, the four anteriors of 1-year-old calves are readily available, are similar to human teeth in size and texture, and are easily mounted for testing.

F. Test Methods

An industrial chemist in need of a standardized, well-proven test method can turn to his ASTM test method books, his Military Specification file, or his trade association file. Unfortunately, no standardized test has yet been developed in the new and rapidly advancing field of adhesive restorations, and each investigator must develop a method while testing his formulation and adhesion factors. Thus, researchers in this field have not been able to compare data readily, and often have difficulty in assessing the significance of their own data because of the interplay of tensile, peel, and flexural forces in the test device and jigs.

Some of the test methods in use include shear (Chang and Uy, 1965) and punch/shear (Patrick, 1965) testing as well as numerous tensile test methods (Phillips and Swartz, 1955; Bowen, 1965; Lee *et al*, 1969; Roydhouse and Rambasek, 1968). All of these methods, however, have one thing in common. They are destructive test methods conducted *in vitro* and at best can only approximate the conditions exist-

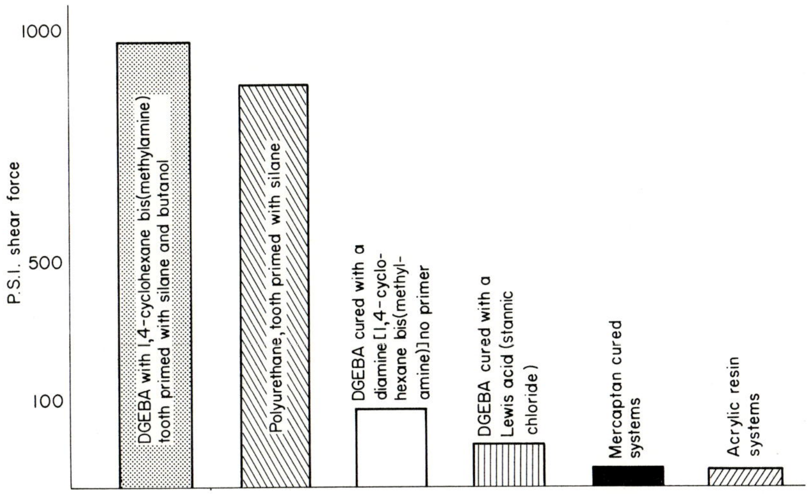

Fig. 9. Adhesion to tooth structure of various resins and primers 24 hrs in 37°C water.

ing in the biological environment, of which all too little is known. At present, there is little other than long-term, *in vivo* clinical testing on a population sample that can yield the data necessary for a complete evaluation of a possible adhesive.

G. Potential Resinous Adhesives to Hard Tissue

There are over 50 major types of resins produced in 100 million or billion pound quantities per annum in the United States. However, despite this vast production and the large investment in plant equipment and in background scientific information, only a very limited number of these resins are usable in medicine or dentistry, and even fewer are potentially adhesive to hard tissue. The field of resins is quickly narrowed by the biological requirements of (1) rapid cure at body temperatures (3–10 minutes), (2) low toxicity, and (3) the temperature range tolerated by vital tissues at which the unpolymerized monomer should be liquid. The choice is then generally limited to one or more of the three types of resins currently being tested as adhesives to hard tissues: Acrylics, including methylmethacrylate, the new thermosetting types, and alkyl cyanoacrylates; epoxies; and polyurethanes.

While the acrylics based on methylmethacrylate have excellent reactivity at body temperature and have been one of the mainstays in medicine and dentistry, they are not inherently adhesive. Because the newer thermosetting acrylics are useful in dentistry, much research has focused on developing adhesion-promoting primers for them. Bowen (1969) has had some success using a chelating agent, the adduct of *N*-phenyl glycine and glycidyl methacrylate. Others (Galligan *et al.*, 1968; Lee *et al.*, 1968) are using polyurethanes to bond the acrylics to the tooth structure.

The epoxies are natural candidates for an adhesive to hard tissue as they develop the strongest and most permanent bonds of any known resin type. Their reactivity, however, is borderline for biological applications, as shown in the Arrhenius plot of reaction rates of several epoxy systems (Fig. 10).

The principal liquid epoxy resin of commerce is the diglycidyl ether of bisphenol A, abbreviated DGEBA. Its structure is shown in Fig. 11. The popularity of this resin stems from a fortuitous set of circumstances. Bisphenol A is the lowest priced diphenol available. Its two benzene rings provide a rigid, tough backbone. These factors coupled with those attributed to glycidyl ethers such as low liquid

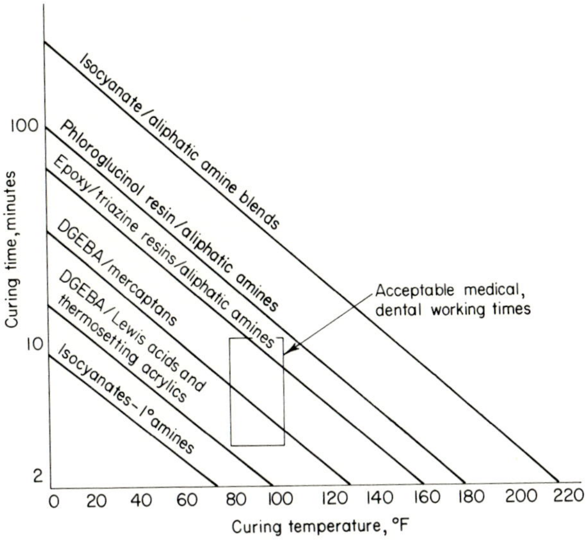

Fig. 10. Arrhenius plot showing reaction rates of small masses of various resins cured with amines, mercaptans, and Lewis acids compared with time and temperature requirements for most operative medical and dental applications.

viscosity, ease of cure with a variety of curing agents, low shrinkage during cure, and cure to a hard, tough, highly adhesive solid, have led to its widespread industrial use.

The basic problem of reactivity of this and other epoxy resins may

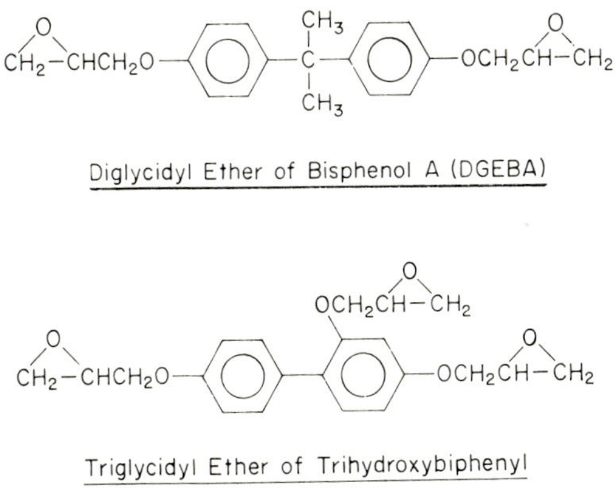

Fig. 11. Examples of two epoxy structures.

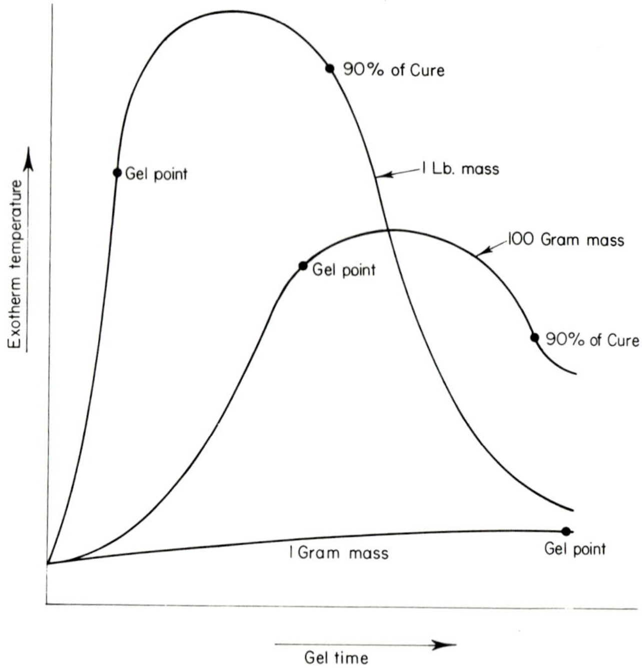

Fig. 12. General effect of mass on exothermic temperature and gel time of epoxy resin.

be separated into three aspects: (1) Influence of mass; (2) effect of temperature; and (3) effect of chemical structure.

The influence of mass on setting time is illustrated in Fig. 12. Here the effect of exothermic heat is illustrated. When large masses are cured, the exothermic heat released is trapped within the bulk of the resin. This heats the resin and accelerates the cure. If the size of the reacting mass is decreased, the exothermic heat has more chance of being conducted away, and the gel time and cure time of the resin are significantly lengthened as shown. Thus, a resin which is used in industrial size masses of 1-19 pounds may set in 10 minutes with a certain amine curing agent; when the mass is reduced to 1-10 grams, it may require hours to set.

The effect of temperature on the reactivity of epoxy resins follows the Arrhenius equation in which a 10°C temperature rise is said to double the rate of reaction. Since in biological applications the maximum temperature is 37°C and the mass of resin used is usually less than 10 grams, the only remaining factor for adjusting reactivity is the chemical structure.

A trifunctional epoxy resin, triglycidyl ether of trihydroxybiphenyl (Fig. 11), was found to be one of the more reactive resins which exhibited good adhesion to hard tooth structure.

III. Epoxy Curing Agents

The carboxylic acids and acid anhydrides have been classified as unsuitable because even with the most powerful accelerators (tertiary amines and tin salts), full cures in small masses have not been achieved under dental clinical conditions. The Lewis bases are generally slow in reactivity, but their investigation with diisocyanate and mercaptans have shown them to accelerate these types of systems.

The primary amines, particularly various alicyclic diamines, appear to have the best general properties as curing agents for dental adhesives but have shown slow reaction. As a result, it is necessary to preheat the resin and amine in order to achieve the desired rapid cure. The system of preheating the epoxy and cure agent works well provided the mass is 5 grams or more. In smaller quantities the initial curing momentum is quickly lost and the cure is prolonged. In biological systems, the ever present moisture necessitates that the resin cure completely within the first 5–10 minutes; otherwise there is a rapid decrease in adhesion as shown in Table I.

The Lewis acids have been found to work somewhat erratically.

TABLE I

Tooth Adhesion Tensile Testing of an Epoxy (Triglycidyl Ether of Trihydroxy Biphenyl) under Various Cure Conditions

Cure conditions			Average tensile force (lb/sq inch)
18 hrs	37°C	95% RH	313
72 hrs	37°C	Water	
10 min	80°C	Oven	322
18 hrs	37°C	95% RH	
72 hrs	37°C	Water	
72 hrs	25°C	Dry	442
1 hr	37°C	Oven	665
72 hrs	25°C	Dry	
15 min	80°C	Oven	988
72 hr	25°C	Dry	

Some of them have suitable cure speeds but their action is inhibited by moist and basic tooth surfaces. A list of some of the Lewis acids tested for adhesion to teeth is given in Table II.

Certain polyfunctional mercapto esters, particularly those based on thioglycolic acid, when accelerated with Lewis bases (tertiary amines), have high reactivity with epoxy resins of the glycidyl ether type, but do not cure epoxy resins of the peracetic acid type. These mercaptoesters, however, have extremely high water absorption and poor ultraviolet resistance, and are flexible at room temperature. They also softened at temperatures as low as 100°–125°F, indicating that they were not fully cured and were still somewhat thermoplastic.

Adhesion of Epoxy Systems to Hard Tissue

Most epoxy resins, such as commercial adhesives, as well as specialized industrial systems noted for being able to cure under water, develop bonds to moist tooth surfaces but lose adhesion after overnight immersion in tap water. Loss of adhesion generally occurs in less than 24 hours. Since diffusion rates through epoxy resins are some 30–40 times lower than for acrylics, our data indicate that moisture is coming to the interface from within the tooth or along the adhesive interface. Various specialized epoxy resins synthesized for biological applications showed retention of adhesion after 36 months in distilled water, when the resins were cured to the tooth with some heat (about 10 minutes at 150°F).

Better adhesion was obtained with the faster, more water resistant systems which had been cured with heat and which were applied to teeth treated with organofunctional silanes. These silanes are capable of reacting chemically both at the surface of hard tissue and with the resin, so that positive primary bonds exist between the tooth and the filling resin. Such chemical bonds would possess bond energies of 20–25 kcal/mole as compared to 6–8 kcal/mole for adsorbed water bonds and 3–4 kcal/mole for adsorbed adhesive bonds. Thus, thermodynamically, the silane treated tissue would be expected to maintain bond strength in the presence of water, whereas adhesive bonds would not.

In summary, epoxy resins in the small masses encountered in biological applications are borderline with respect to having the necessary rapid cure. Those epoxy systems which can be made to set rapidly (usually within one minute) will not bond adequately. While epoxies have exhibited good adhesion to hard tissue with adequate heat cure, they are not at present clinically feasible. It is probable, however, that better accelerators can be developed.

TABLE II
LEWIS ACIDS INVESTIGATED FOR ADHESION TO HARD TISSUE

$AlCl_3$	$(CH_3)_2SiCl_2$	PO_5
$Al(C_3H_7O)_3$	$CuCl_2$	$POCl_3$
AlF_3	$FeCl_3 \cdot 6H_2O$	$SbCl_3$
$AsCl_3$	$HSiCl_3$	$SiCl_4$
BaCl	H_3BO_3	$SnCl_2$
$BF_3 \cdot Et_2O$	H_3PO_4	$SnCl_4 \cdot 5H_2O$
$BF_3 \cdot MeOH$	$HgCl_2$	$SOCl_2$
$BiCl_3$	$ØSnCl_2$	SO_2Cl_2
$BuSnCl_3$	$Hg(NO_3)_2 \cdot 1/2H_2O$	$TiCl_3$
Bu_2SnCl_2	$Hg_2(NO_3)_2 \cdot 2H_2O$	$Ti(OC_4H_9)_4$
$Bu_2Sn(OCOCH_3)_2$	$LaCl_3$	$UO_2(C_2H_3O_2)_2 \cdot 2H_2O$
CaF_2	$La(NO_3)_3$	$UO_2(NO_3)_2 \cdot 6H_2O$
$CdCl_2$	LiF_6Si	$ZnCl_2$
$Cd(NO_3)_2 \cdot 4H_2O$	$MgCl_2$	$Zn(C_2H_3O_2)_2$
$CH_2{=}CHSiCl_3$	$MnSO_4$	$Zn(BF_4)_2$
CH_3SiCl_3	NH_2SO_3H	$ZrOCL_2 \cdot 8H_2O$
$CH_3CH_2SiCl_3$	PCl_3	ØSiCl3
CH_3SiHCl_2		

Acknowledgments

The work described in this paper was performed under contract to the National Institutes of Health, National Institute of Dental Research, under Contract PH 43-64-548.

References

Bowen, R. L. (1963). *J. Am. Dental Assoc.* **66**, 37–64.
Bowen, R. L. (1965). *J. Dental Res.* **44**, 895–911.
Bowen, R. L. (1967). *J. Am. Dental Assoc.* **74**, 439.
Bowen, R. L. (1969). *46th Gen. Meeting Intern. Assoc. Dental Res., 1968.* (to be published).
Chang, R., and Uy, K. C. (1965). *U.S., Public Health Serv., Publ.* **1494**.
Galligan, J. D., Schwartz, A. M., and Minor, F. W. (1968). *J. Dental Res.* **47**, 629–632.
Lee, H. L., Swartz, M. L., and Cupples, A. L. (1968). *Ann. Comp. Rept. Natl. Inst. Dental Res., 1968* Epoxylite Corp. RR 68–140. Clearinghouse, Natl. Bur. Std.
Lee, H. L., Swartz, M. L., and Culp, G. (1969). *J. Dental Res.* **48**, 211–216.
Patrick, R. L. (1965). *U.S., Public Health Serv.,* **1494**.
Phillips, R. W., and Swartz, M. L. (1955). *J. Am. Dental Assoc.* **59**, 172–177.
Provenza, D. V., and Sardana, R. C. (1965). *U.S., Public Health Serv., Publ.* **1494**.
Roydhouse, R.H., and Rambasek, G. M. (1968). *J. Dental Res.* **47**, 1087–1093.
Smith, D. L., and Schoonover, I. C. (1953). *J. Am. Dental Assoc.* **46**, 540–544.
Zisman, W. A. (1965). *U.S. Public Health Serv., Publ.* **1494**.

Author Index

Numbers in italics refer to the pages on which the complete references are listed.

N

O

P

Q

R

Subject Index